Recent Developments in Manufacturing Robotic Systems and Automation

Editors

Dan Zhang and Zhen Gao

Robotics and Automation Laboratory
Faculty of Engineering and Applied Science
University of Ontario Institute of Technology
2000 Simcoe Street North
Canada

Bentham Science Publishers
Executive Suite Y - 2
PO Box 7917, Saif Zone
Sharjah, U.A.E.
subscriptions@benthamscience.org

Bentham Science Publishers
P.O. Box 446
Oak Park, IL 60301-0446
USA
subscriptions@benthamscience.org

Bentham Science Publishers
P.O. Box 294
1400 AG Bussum
THE NETHERLANDS
subscriptions@benthamscience.org

CONTENTS

FOREWORD

Robotic and Automation systems are the core of today achievements towards future developments of Science and Technology.

But challenges are still faced in many aspects for a full success of those systems in industrial and non-industrial applications for which fundamental are design and manufacturing issues.

This eBook presents a collections of contributions from authors all around the world as a fact of interest in those aspects in all countries around the world, although with different perspectives. The book editors Dan Zhang and Zhen Gao have succeeded in inviting authors for different aspects, but with the challenging idea of asking contributions to young and emerging researchers and designers, who are active actors in the field. The absence of very famous names among contributors gives room to fresh blood for new ideas and proposals in a field where a huge literature can show an impression of saturation in discussions for novelty.

Innovation and advances in the fields of Robotics and Automation are nowadays continuously proposed with characters of novelty and update of past solutions. This eBook is a very good expression of those challenges, namely contributions by new actors in the field and innovative advances for future developments.

Surely, a reader will get satisfaction from this eBook, but she/he will also get inspiration for her/his work and further contribution in those and other aspects of Robotic and Automation Systems.

Marco Ceccarelli

Laboratory of Robotics and Mechatronics
DICeM; University of Cassino and South Latium
Via Di Biasio 43, 03043 Cassino (Fr)
Italy

PREFACE

The emergence of robotics technologies has made a significant contribution in modern manufacturing industry. Because of the current trend toward high-precision and high-speed machining, there is an increasing demand to develop manufacturing robotic system with improved performances including higher stiffness, higher dexterity, higher adaptability, and higher reconfigurability.

Current research activities employ the technologies of sustainable manufacturing to advance and promote the development of modern manufacturing. In recent years, the need for system sustainability and a globally increasing manufacturing, drive a gigantic demand for technology and strategies which will reduce production costs. It is expected that the advanced robotics and automation technologies will revolutionize the entire manufacturing section in the near future.

This eBook includes the start-of-the-art robotic systems and automation for advanced manufacturing, and introduces the latest development of design methodology, kinematics and dynamics analysis, performance analysis and evaluation, intelligent manufacturing, assembly, sensors, control theory and practice, human-machine interface, and so on.

This eBook consists of seven chapters. Chapter 1 introduces a novel design method composed of suitable phases that generate feasible mechanical designs of robotic systems. Chapter 2 presents the main solutions to design order picking automation systems. Chapter 3 provides a visual one to one comparison between G & M Code (ISO 6983) and STEP-NC Code (ISO 14649) of computerised numerical control issues. Chapter 4 discusses the characterization of reconfigurable Stewart platform for contour generation and active vibration isolation applications. Chapter 5 presents an example of 2-DOF global kinematic model generation to demonstrate the methodology and application of reconfigurable modeling theory. Chapter 6 reviews the updated technologies that are used to achieve an assembly environment with high flexibility for the cognitive assembly systems. Chapter 7 presents the kinematics analysis and the teleoperation system of a 4-DOF modular reconfigurable robot.

Finally, the editors would like to sincerely acknowledge all the friends and colleagues who have contributed to this eBook. Special thanks to Ms. Maria Baig. Without your great assistance, this eBook cannot be published.

Dan Zhang and Zhen Gao

Robotics and Automation Laboratory
Faculty of Engineering and Applied Science
University of Ontario Institute of Technology
2000 Simcoe Street North
Canada

List of Contributors

Ana Djuric Intelligent Manufacturing Systems Centre, University of Windsor, Windsor, Canada

Waguih ElMaraghy Intelligent Manufacturing Systems Centre, University of Windsor, Windsor, Canada

Zhen Gao Faculty of Engineering and Applied Science, University of Ontario Institute of Technology, Ontario, Canada

Noordiana Kassim Faculty of Mechanical and Manufacturing Engineering, Universiti Tun Hussein Onn, Malaysia

Jianhe Lei Faculty of Engineering and Applied Science, University of Ontario, Institute of Technology, Ontario, Canada

Marco Melacini Department of Management, Economics and Industrial Engineering, Polytechnic University of Milan, Itlay

Nagarajan Thirumalaiswamy Department of Mechanical Engineering, Universiti Teknologi Petronas, Perak, Malaysia

Li-Ming Ou Department of Mechanical Engineering, University of Auckland, Auckland, New Zealand

Nestor E. Nava Rodríguez Robotics Lab, Carlos III University, Calle Universidad, Madrid, Spain

Kumar G. Satheesh Department of Mechanical Engineering, Indian Institute of Technology Madras, India

Zhanglei Song Faculty of Engineering and Applied Science, University of Ontario, Institute of Technology, Ontario, Canda

Xun Xu Department of Mechanical Engineering, University of Auckland, Auckland, New Zealand

Yusri Yusof Faculty of Mechanical and Manufacturing Engineering, Universiti Tun Hussein Onn, Malaysia

Dan Zhang Faculty of Engineering and Applied Science, University of Ontario, Institute of Technology, Ontario, Canada

2

CHAPTER 1

A Novel Design Approach for Robotic Systems

Nestor E. Nava Rodriguez[*]

Robotics Lab., Carlos III University, Calle Universidad, 30, 28911 Leganes (Madrid), Spain

Abstract: The mechanical design is composed of certain stages that are defined by every designer following his/her personal method and experiences. The work can be divided as much as necessary from the study of system requirements to the manufactory. The design quality depends of the work developed in every stage and the number of these steps provides the level of design accuracy. The results of design process with enough effective stages are accurate and fulfil the requirements that have been built for. The stage number, which can be defined as work to develop, should be just the necessary for a successful design. This chapter presents a novel design method composed of suitable phases that generate feasible mechanical designs of robotic systems. Examples of robotic systems designed by applying this method have been illustrated. The experimental tests of these systems operation illustrate successful results that validate the effectiveness of the proposed approach.

Keywords: Systems, robotics, robotic mechanisms, mechanism examples, design, novel design, design process, suitable design, approach, effective approach.

INTRODUCTION

The approach adopted by [1] for defining design is viewed as the total activity necessary to provide a product o process to meet a market needing. The mechanical design involves skills in the mechanism and components area, such as bearings, shaft, gears, seals, belt, chain drivers, clutched, brakes, spring, fasteners and so on. Most of these components are available in the market and the necessary step for their specification and selection represent an important key in the design process. Robots are complex machines that require special efforts for the design process. Usually they are complex systems, which design and construction are a

---

*Address correspondence to Nestor E. Nava Rodriguez:** Robotics Lab., Carlos III University, Calle Universidad, 30, 28911 Leganes (Madrid), Spain; Tel: +34 916248813; Fax: + 34 91 624 9339; E-mail: nnava@arquimea.com

challenge for researchers in several fields of science and engineering. At least the mechanical functionality of main components has to be clearly understood in order to design the basic mechanical features. In most of cases, the cost of robots manufacturing is as high as level of complexity. Low-cost robots are related to new emerging application areas [2]. In fact, increasing robot utilization may be viewed in term of low-cost design and operations. A low-cost robot can be conceived by using commercial components with a robust and no so-complex mechanical design as well as easy-operation through flexible programming. Nevertheless, low-cost robots provide mechanical structures with limitations in terms of performance versatility. Therefore, the low-cost level in robot design should be considered for applications in which obtained solutions provide successful operations.

Simulation is a useful tool for modelling engineering systems that involve several operations. Traditional simulations are composed of mathematical algorithms that can be resolved by computer assisting. Performances of systems under different conditions can be evaluated by using simulations. They are able to predict the system behaviour from a set of design parameters and constraints. Therefore, simulations development can be a feasible approach in the validation of mechanical system before construction.

This chapter deals with a new approach for the design of robotic systems with an accurate mechanical structure. The framework used within this chapter has been to provide descriptive and illustrative information to introduce principles to expose the reader the detailed methods and calculation necessary to specify, design or select suitable components. The aim is to provide the reader with sufficient information to develop necessary skills to carry out calculations and selection process. The proposed approach is composed of effective and enough stages that involve the most important topics of the mechanical design, such as kinematics, dynamics, parts design, validation, testing and optimization. The risk of elaborate unneeded work or repeat what is done can be run by carrying out too many stages. Nevertheless, only few stages can provide a non accurate solution. The initial two phases represent the mechanism synthesis giving by the kinematic and dynamic design. The third stage is the mechanical design of parts and components selection that is carried out by using CAD tools. Afterwards, computer validation is proposed by study results of kinematic and dynamic simulations of operations, elaborated in a virtual environment, of a virtual system. Kinematic and dynamic simulations can be also used in the mechanism synthesis stages since they represent powerful tools that help to compute complex formulations. Feedback from the computer assisted validation inputs the initial three stages in order to correct some mistakes yielded during the mechanical design. Once the computer validation shows successful results of operating simulation, the manufacturing can be launch that represents the fifth stage of the proposed approach. The sixth phase is an experimental test on a built prototype that is useful to study of operation of a real structure. In some cases, mainly in complex systems, the simulators do not allow including characteristics for a working environment enough close to the reality. Therefore, the experimental experiences are necessary to check the results of the computer assisted validation. Finally, an optimization process can be elaborated in order to enhance the design even more. Detailed examples and worked solutions are supplied throughout the manuscript.

A NOVEL APPROACH FOR ROBOT DESIGN

The robot design represents a complex process that requires a careful development of specific issues providing successful systems with suitable characteristics. Fig. **1** shows a flow chart with the stages proposed for the new design approach and their sequence of application. The first part of the approach represents a conceptual design that is composed of three stages, such as kinematic, dynamic and mechanical design. Afterwards, the prototyping and validation of built devices can be carried out just after computer simulations that provide a model validation. This strategy avoids the prototype manufacturing with unfeasible characteristics. Every design result can be improved even after obtaining a final solution, thus an optimization process can be elaborated. Feedback to every stage of the conceptual design is prevented in order to correct any problem detected during prototyping and validations. Similar, the conceptual design can be part of the optimization or some aspects can be optimized and then inputted to the design process, as shown in Fig. **1**. The proposed design method is an iterative process in which the number of iterations depends of the quality of work developed, so as long as the design is well done the iterations are less. Other factor to taking into account is feedbacks from experts in other fields to mechanical designers in order to prevent possible problems provided from the chosen solutions.

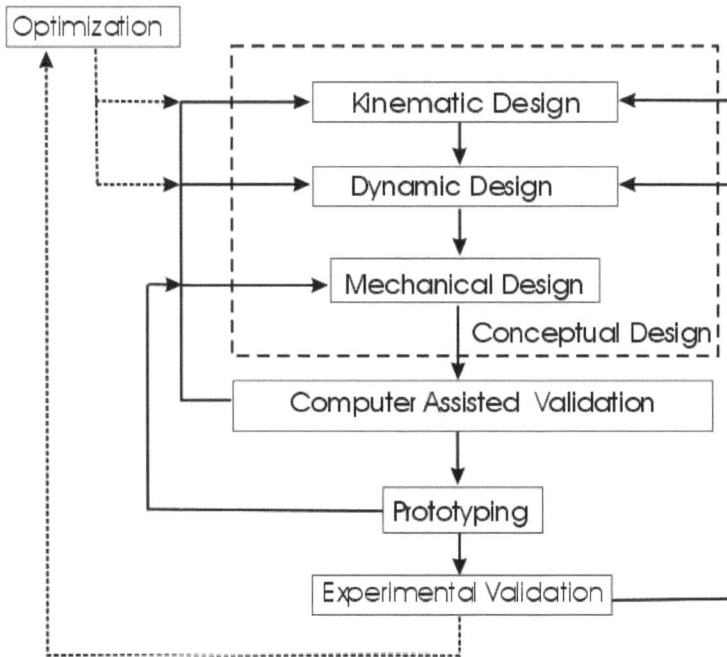

Figure 1: Flow chart of the proposed design approach.

In fact, assumed design results can be suitable from the mechanical point of view but they can also be unfeasible for specific systems, such as vision or cognitive, as reported below in some illustrative examples. Using the assistance of computer software for both design and validation is an important characteristic of the proposed strategy. These tools allow the elaboration of realistic virtual models as well as simulate operations, reducing computation time, without a built system, and saving cost. Now, every phase of the proposed design approach is illustrated through significant examples of developed robotic systems.

Kinematic Design

The aim of this phase is to design the movement characteristics of the robot for a proper operation in specified application. The kinematic chain is studied in terms

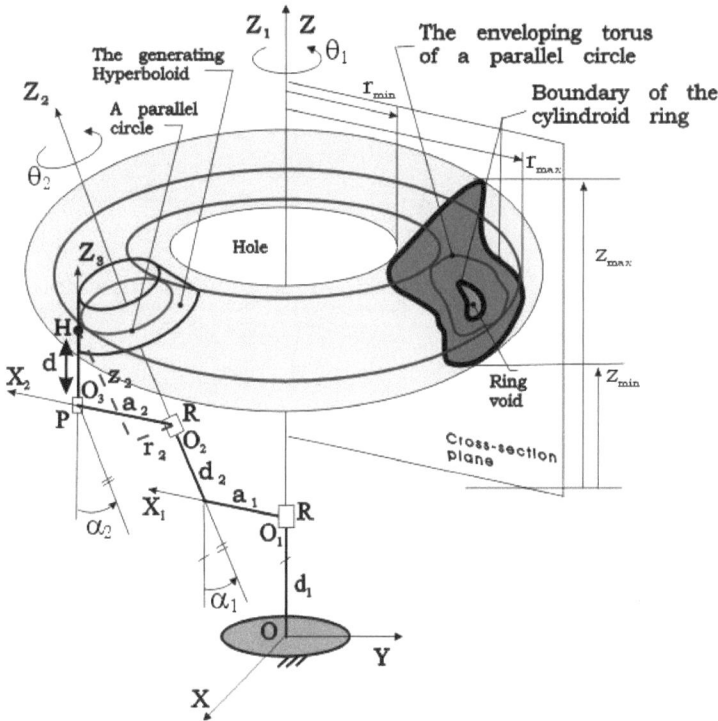

(a)

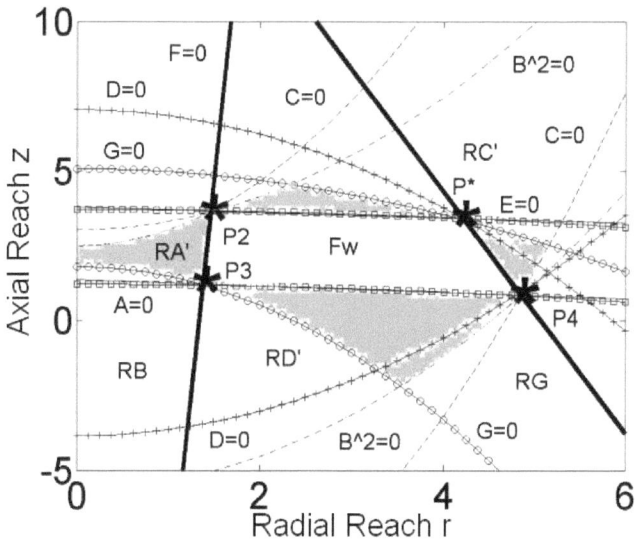

(b)

Figure 2: Workspace analysis of serial manipulators: (a) workspace shape of a telescopic manipulator [3]; (b) feasible workspace region of a general R-R-R manipulator [4].

of configuration, link dimensions and numbers of degrees of freedom (DOF) in order to obtain the desired position, velocity and acceleration for both end-effector and joints. Fig. **2(a)** shows an example of a kinematic study that has been carried out for the design of a telescopic manipulator. An algebraic formulation has been presented for representation of the boundary workspace of RRP manipulators in Cartesian Space [3]. The cross-section boundary curve and boundary surface are of 8-th degree. Special cases have been studied in order to illustrate all possible simplifications of the expression for the cross-section boundary curve. Moreover, the effect of twist angles α_1 and α_2 as well as links parameters a_1, a_2 and d_2 on the workspace of RRP manipulators have been considered by analyzing special cases and their combinations. Depending of the D-H parameters the RRP workspace boundary has a specific shape. Therefore, a telescopic manipulator can be designed by considering a desired workspace region inside the shapes obtained in [3]. The kinematic parameters of the designed manipulator can be obtained from its workspace shape. Similar, an analytic procedure has been proposed for computing the feasible workspace region of a generic R-R-R serial manipulator [4]. A formulation based on the ring equation has been used for computing expressions of structural coefficients of a manipulator. The determination of the feasible workspace region of Fig. **2(b)** (in grey) can be useful both in the design and analysis of a manipulator with specific workspace characteristics.

The structural equations have been restricted analytically in order to define sub-regions that compose the feasible workspace region. The feasible sub-regions can be obtained by prescribing four points P*, P2, P3 and P4 of the manipulator workspace boundary. These points can be desired points of the end-effector trajectory of a designed manipulator. Therefore, the obtained manipulator has a workspace boundary inside the grey region of Fig. **2(b)** and its kinematic parameters can be obtained by computing the formulation in [4].

Fig. **3** shows the kinematic design of an articulated robotic finger with 1 DOF. The results of experimental tests on human hands have been used in order to define its kinematic characteristics [5]. This finger has the same size of the average human index finger. The kinematic architecture of Fig. **3(a)** permits to

(a)

(b)

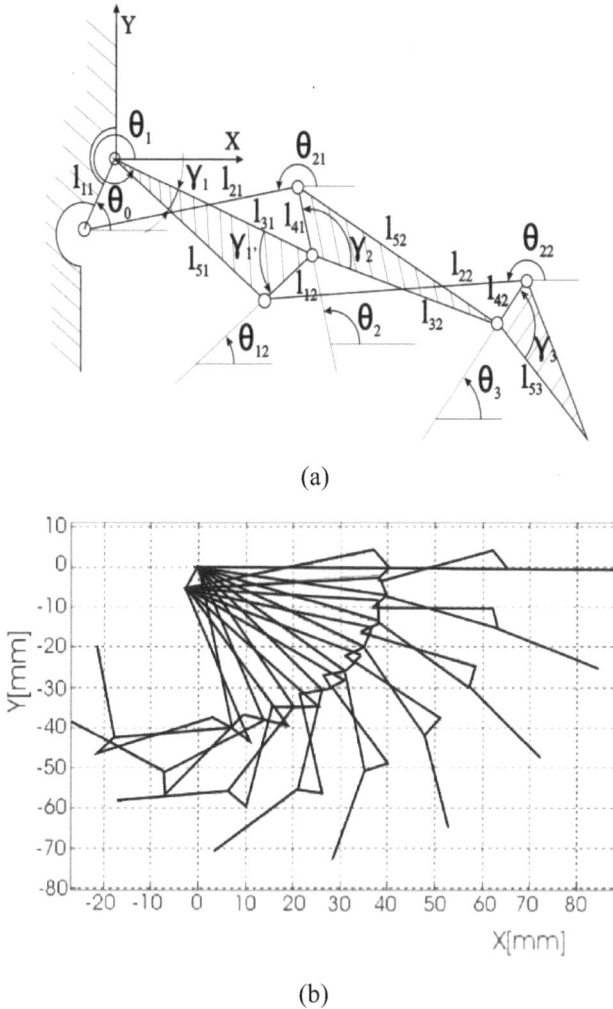

Figure 3: A 1-DOF articulated mechanical finger [5]: (a) a kinematic scheme; (b) simulation of the finger performance.

mimic the movement of human fingers with only 1 DOF. It has been decided to actuate the first phalanx by connecting it directly to a motor or through a proper gear train. The other phalanxes are actuated by an articulated mechanism that is connected with the previous phalanx. In order to obtain a robust mechanical design with suitable stiff, it has been decided to use an articulated mechanism for driving the finger prototype. The drive-finger mechanism is a series of crossed-four bar linkages mechanism. The mechanism is robust and it can be embedded in the finger body. Nevertheless, special care has been addressed to the dimensional synthesis of the articulated mechanisms in order to obtain the needed transmission ratios, to avoid link interference and to achieve suitable mobility. Fig. **3(b)** shows a simulation of the finger operation that has been used in order to check performance with the desired kinematic characteristics. This design has been studied and solution has been obtained with satisfactory results by looking at the different aspects separately and by achieving a dimensional synthesis of the mechanism with traditional techniques [5].

By using previous experiences with a test-bed of a pneumatic gripper and system, a design of a new pneumatic gripper with two-fingers has been developed with low-cost easy-operation features [6], as shown in Fig. **4**. The new gripper uses an articulate mechanism in order to drive the movement of the fingers. The driving finger mechanism of the gripper is a Withworth quick return mechanism, [7]. Including this driving mechanism to the gripper is low-cost and easy-operation solution and provides robustness to the design. The mechanical design of the new gripper prototype has been based on the kinematics model of Fig. **4**, in which F indicates the pneumatic actuator force; l indicates the actuator displacement; d indicates the first link of mechanism; b indicates the second link; a indicates the third link; c indicates the fourth link; e indicates the end-effector displacement. β indicates the angle of link a and ϑ indicates the angle of link c with respect to X reference axis.

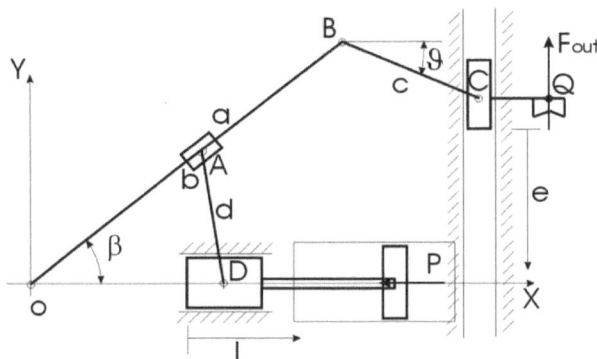

Figure 4: Kinematic scheme of a pneumatic actuated gripper [6].

Fig. **5** shows kinematic simulations for the design of leg and arm sub-systems for a low-cost humanoid robot called CALUMA (CAssino Low-cost hUMAnoid

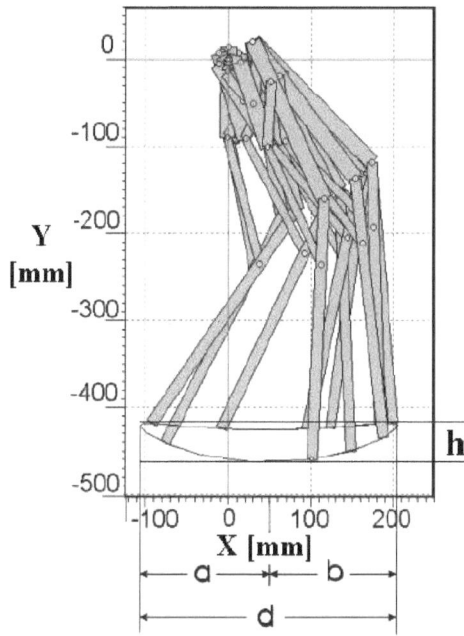

(a)

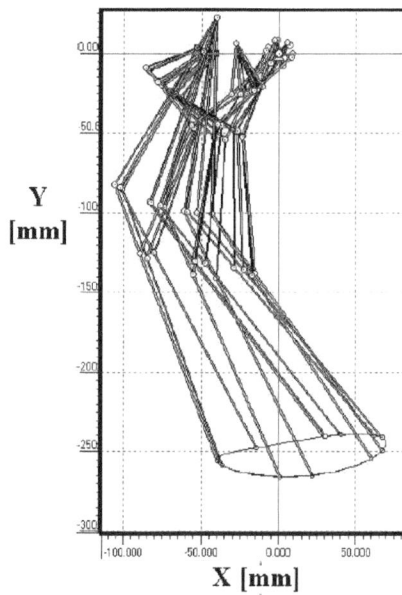

(b)

Figure 5: Kinematic simulations for the design of a low-cost humanoid robot [8]: (a) legs sub-system performance; (b) arm sub-system performance.

robot) [8]. In particular, Fig. **5(a)** shows the performance of a leg mechanism that represents a biped robot with only one actuator for both legs. This structure uses a Chebychev-Pantograph mechanism in order to transmit the movement to the feet, as illustrated in [9]. This leg-driving mechanism has been modified in order to obtain a suitable performance of legs for the equilibrium of the whole humanoid structure during walking. In Fig. **5(a)**, the structure of the new leg-driving mechanism can be recognized. The trajectory of the end-effector is indicated in Fig. **5(a)** with a continuous line. The new structure of the leg module permits obtaining symmetrical steps with smaller dimensions than the initial solution [8]. Similar, Fig. **5(b)** shows an arm sub-system for CALUMA, which uses the same transmitting mechanism of the leg modules for driving the end-effector but with suitable arm characteristics. Fig. **5(b)** shows a simulation of the arm mechanism indicating the trajectory of the end-effector. The arm module has the same structure of leg but in an inverted configuration. The movement of arm links is opposite respect to the leg movement, as shown in Fig. **5**. The dimensions of arm structure are 1.618 smaller than leg structure. The number 1.618 is called "The Divine Proportion" or number φ. This number is also the proportion between human arms and legs [10] and it has been used in order to obtain an anthropomorphic design of CALUMA [8].

The kinematic design provides the performance of the mechanical structure as well as information for the dynamic design development, such as joint position, velocity and acceleration.

Dynamic Design

By using the results of the kinematic design, the dynamic design can be carried out in order to obtain the dynamic parameters of the system, such as forces, torques, mass, inertia and friction coefficients. For example, in the design of the robotic finger of Fig. **3**, a force analysis has been carried out in order to obtain the forces that each phalanx can apply [5] and compute de actuating torque C_m. It has been assumed that the resultant of gasping force on each phalanx is applied in the middle of the phalanx itself. By using this assumption one can write the principle of virtual works as

$$C_m \dot{\theta}_1 = F_1 V_1 + F_2 V_2 + F_3 V_3 \qquad (1)$$

Since the mechanical finger has been designed considering anthropomorphic characteristics, the values of F_i (with i = 1 to 3) can be assumed that obtained from measuring the human grasping force. The maximum values of forces can be considered for computing Eq. (1) in order to assume the worse case. The value of the velocity component V_i of the i-th (with i = 1 to 3) contact point can be obtained by using the expressions [11]

$$^{i+1}\omega_{i+1} = {}_{i+1}R_i\,{}^i\omega_i + \dot{\theta}_{i+1}\,{}^{i+1}\hat{Z}_{i+1}$$
$$^{i+1}v_{i+1} = {}_{i+1}R_i\left({}^i v_i + {}^i\omega_i + {}^i S_{i+1}\right)$$

$$(2)$$

where $^{i+1}\omega_{i+1}$ is the angular velocity of the i-th phalanx with respect to the frame of (i+1)-th phalanx; $_{i+1}R_i$ is the rotation matrix; $^i\omega_i$ is the angular velocity of the i-th phalanx with respect to the frame of i-th phalanx; θ_{i+1} dot is the angular velocity of phalanx joint (i+1)-th and $^{i+1}Z_{i+1}$ is its direction along Z_{i+1} phalanx axis; $^{i+1}v_{i+1}$ is the linear velocity of the (i+1)-th phalanx with respect to the frame of (i+1)-th phalanx; $^i v_i$ is the linear velocity of the i-th phalanx with respect to the frame of i-th phalanx; $^i S_{i+1}$ is the distance vector between Z_{i+1} and Z_i phalanxes axes.

Similar, the human-like grasp can be obtained in the finger mechanism by looking at the contacts, in terms of their number and locations, and even in terms of the contact forces that can ensure a static equilibrium of grasped objects. These aspects can be studied through suitable models and using them for both design and operation purposes [12]. Basic computations have been modelled by referring to the planar two-finger grasp of Fig. **6(a)**. One contact for each phalanx has been assumed in the design model of Fig. **6(a)** through contact forces F_{ij} (i=1,2; j=1,2,3). In this model the friction components have been neglected in order to evaluate the grasping efficiency mainly due to the mechanical design of the finger. Angles ϕ_{ij} are the angles between the corresponding forces F_{ij} and velocity V_{ij} of the contact points; and t_i (i=1,2) is the actuation torque on each finger mechanism. A formulation for the static grasp of a human finger can be obtained by using the free-body method for a mechanical design of an anthropomorphic finger in order to study the static equilibrium of phalanxes. Fig. **6(a)** shows a general model for the static equilibrium of the grasped object that has been used to derive the static equilibrium equations. Since the palm plays a key role in the static grasp of a human finger, in the model of Fig. **6(a)** the reaction force $\mathbf{F}_{00}$ of the palm has been taken into account for obtaining the static equilibrium equations.

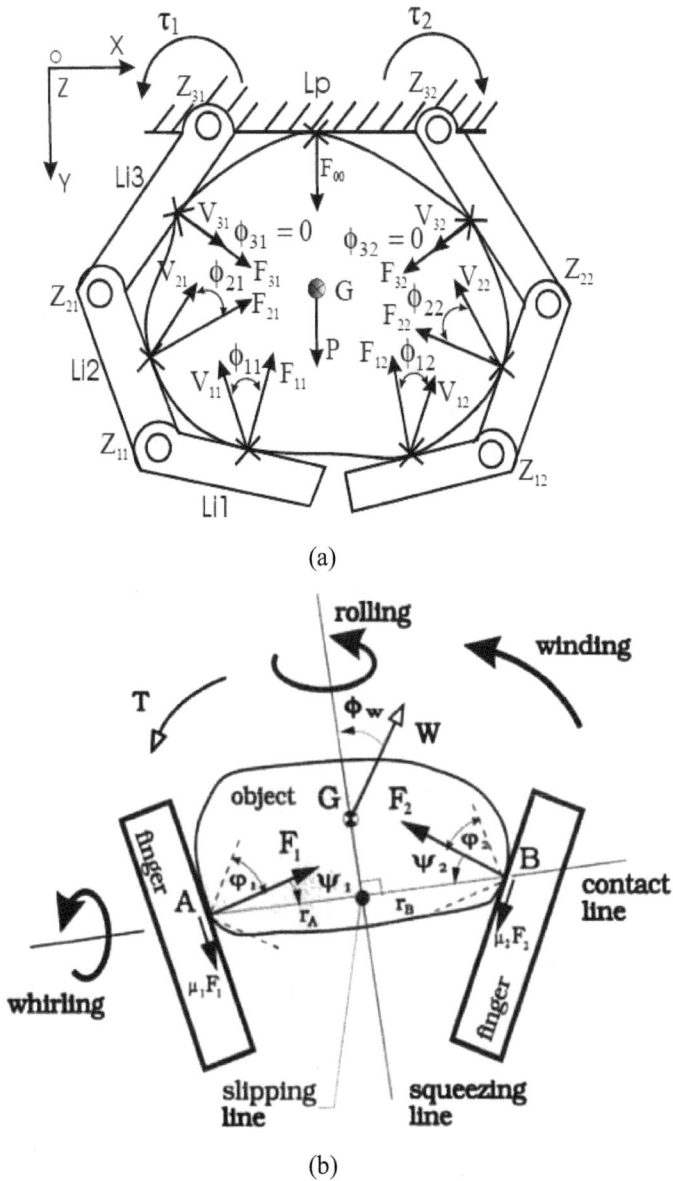

(a)

(b)

Figure 6: Dynamic schemes a 2D study of grasping: (a) gripper grasp [12]; (b) robotic hand grasp [13].

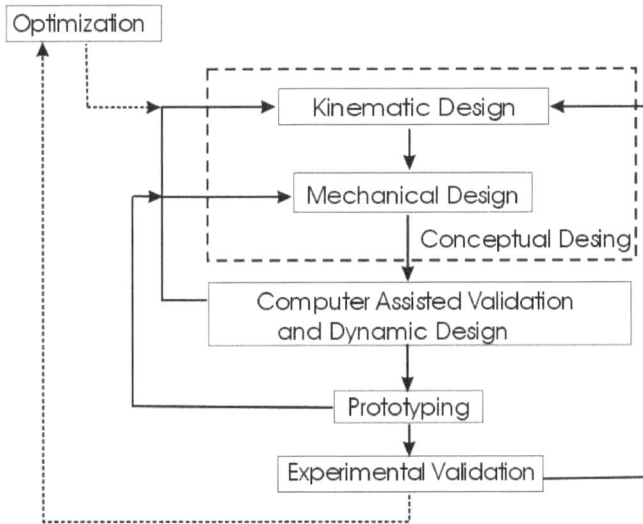

Figure 7: An alternative for the proposed design of Fig. **1**.

Fig. **6(b)** shows a dynamic scheme of a two-fingered gripper grasping an uneven object. The model of Fig. **6(b)** can be used in order to compute the grasping force necessary to perform a firm grasp with the gripper of Fig. **4**. Important characteristics of the gripper grasping can be recognized in these scheme, such as the squeezing line, contact line and rolling, winding and whirling torques [13]. Therefore, the actuating force can be also obtained by computing the dynamic formulation of the gripper mechanism.

Alternative Approach for the Dynamic Design

Since the computer software for dynamic simulations represent a powerful tool that is so-useful for computing the dynamic formulation of complex robotic systems, an alternative approach can be assumed in which the dynamic design is included as part of the computer assisted validation, as shown in Fig. **7**. For example, in the design procedure of the humanoid robot CALUMA, dynamic simulations of the humanoid performance have been developed in order to check the feasibility of its operation. An ADAMS model of CALUMA robot has been developed in order to study the operation, task performances and feasibility of the design. The MSC.ADAMS® is a software that permits to obtain realistic simulations of full-motion behaviour of complex mechanical systems, as outlined in [14]. During this procedure, problems in the kinematics and mechanical design have been recognized and solutions have been fed back to the conceptual design, thus CALUMA robot has presented an evolution of its design, as shown in Fig. **8** and reported in [15]. The first proposed structure for CALUMA has been composed mainly for the original prototypes of limbs and trunk that have been built at LARM (Laboratory of Robotics and Mechatronics of the University of Cassino) [16]. Some modifications can be recognized in the sub-system structure,

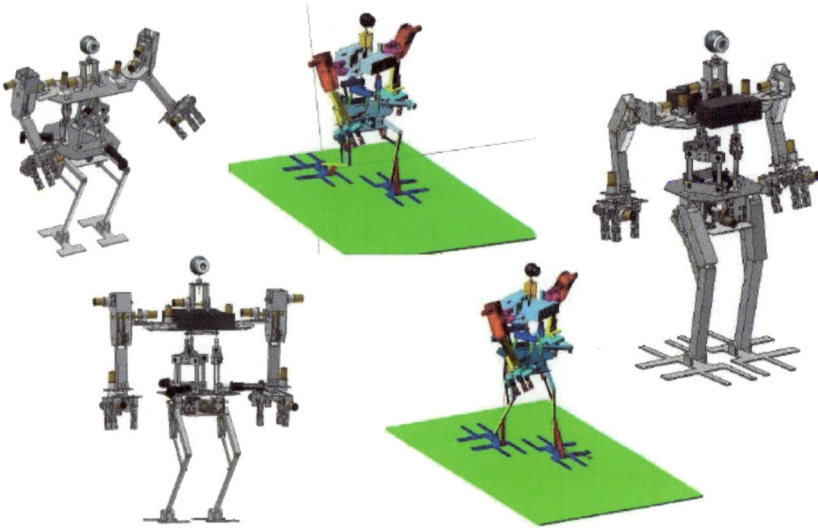

Figure 8: Design evolution of a low-cost humanoid robot [14].

which mainly consists on permitting connection between sub-systems for a whole humanoid assembly. Only the arm sub-system of the first humanoid structure has not been the original LARM prototype. In the second proposed prototype of CALUMA robot, a wrist mechanism and the PLCs Siemens S7 as part of control system are included. Then, a new design of foot has been considered for a third version. Leg structures with a modified Chebychev-Pantograph mechanism has been considered in a new CALUMA proposed model. The fully-right model of Fig. **9** shows final proposed version of CALUMA structure. The leg and arm sub-systems with a modified Chebychev-Pantograph mechanism of Fig. **5** and more proper commercial components are part of this robot structure. The trunk sub-system has a more compact design with suitable position and dimensions of actuators. The proposed robot of Fig. **9** is a practical, efficient and versatile design for a humanoid robot by still using low-cost easy-operation solutions. Its sub-system mechanisms and components give a robust and compact design. The low-number of DOFs provides no-complex actuation that can be operated by the proposed control system. In order to size the leg of CALUMA components, a static analysis of the robot structure has been carried out. Preliminary walking simulations have been used for estimating the applied maximum forces on the structure and the most stressed components.

This simulation has consisted in one step of CALUMA with a velocity of 1 s/step. Fig. **9** shows a time history with the maximum reaction forces that has been obtained during a preliminary walking simulation. In Fig. **10**, the joints with maximum reaction forces are indicated as left joint 3, left joint 4 and left joint 6. Note the maximum value of reaction forces have been obtained at about 0.45 sec of simulation time. At left joint 3 the force has been about 250 N; at left joint 4 the force has been about 225 N; and at left joint 6 the force has been about 200 N. These forces have been used for computing the equations of force and torque that are given by, [17]:

$$V = F_1(X-0)^0 + F_2(X-50)^0 + F_3(X-200)^0$$
$$M = F_1(X-0) + F_2(X-50) + F_3(X-200)$$

(3)

where V is tangent force, M is tangent torque, Fi are reaction forces and X is the position of the studied point.

Fig. **9** shows a scheme of distributions of forces on the most stressed component. The proposed cross-section of Fig. **9** is a cross-section of commercial rods with external dimensions of 28 x 28 mm and internal dimensions 22 x 22 mm. Fig. **10** indicates the points A, B, pure compression and pure traction that have been studied in order to find out the point with highest stress. The stress matrix and the principal stresses I_1, I_2 and I_3, [18], have been computed for the above-mentioned points. By using Von Mises (VM) method, [18], the VM effort has been computed. Point A has been the most stressed point with a VM stress of 13.19 MPa. A commercial aluminum alloy has a Young's modulus about of 100 MPa, [19]. Therefore, CALUMA leg module can be designed by using the above-mentioned commercial rods.

The results of the dynamic design are a first estimation of the torques at the actuating joints for the selection of actuation. Moreover, the forces and torques applied on the structure can be computed for the parts design as well as material and commercial components selection.

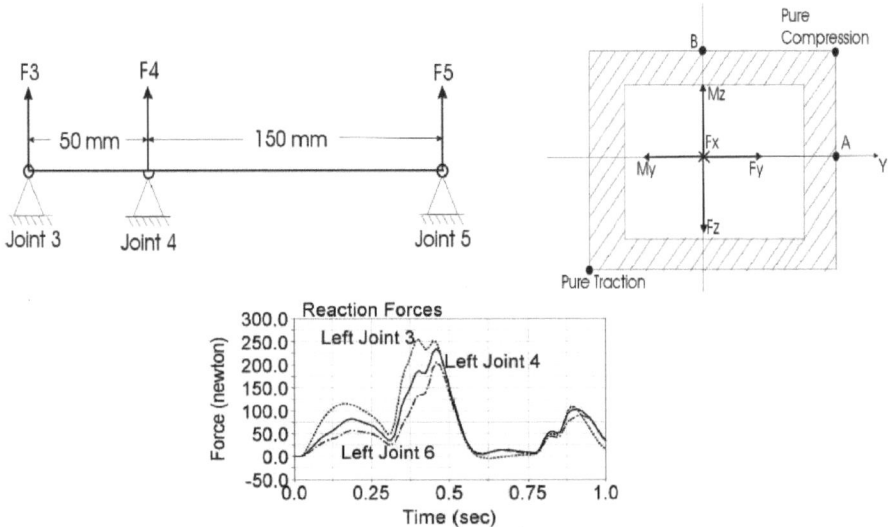

Figure 9: Stress analysis of CALUMA structure by using the dynamic simulation of its operations [8].

Mechanical Design

The mechanical design consists of the modeling of the robot parts and selection of commercial components. The use of CAD tool helps for facilitating and accelerating this task. It is important to emphasize that the pieces design must be carried out thinking of their manufacturing and assembly. Otherwise, the design cost can be high and in some cases so-difficult to construct. Similar, the components assembly should be planed for easy installation that should be considering in the parts design. For example, a three fingers robotic hand has been designed by using the new 1 DOF finger module of Fig. **3**. Each mechanical component of the hand has been designed by using Autodesk Inventor [20], even for the analysis of mechanical interference and to simulate the grasping, before the construction of the prototype. Two hand models have been proposed as based on low-cost and compact design constraints as shown in Fig. **10**. The principal difference between the first model and the second proposed hand is the design of the palm. In fact, the first hand has a more compact and simple model of palm as shown in Fig. **10(a)** and the second hand has a different shape of the palm as shown in Fig. **10(b)**. The shape of the palm of the second hand has a different orientation of the fingers and different location of actuators respect to the first model hand. These aspects make the second design more anthropomorphic. Moreover, the actuation of the second proposed hand can be either pneumatic or electric. Similar, Fig. **11** shows a 3D CAD model of the gripper of Fig. **4** that has been developed in Autodesk Inventor environment. The developed 3D model has been used in order to check the feasibility of assembling all components and checking operation. The gripper design has been proposed as based on low-cost and compact design constraint. For manufacturing the links, commercial aluminum alloy has been used since it is low-cost and lightweight. The proposed actuation for the new gripper prototype is pneumatic.

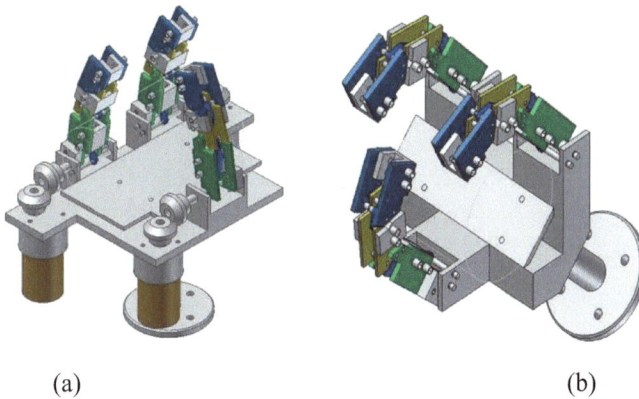

(a)   (b)

Figure 10: Mechanical designs for a robotic hand with the articulated finger of Fig. **3** [5]: (a) First model, (b) Second model.

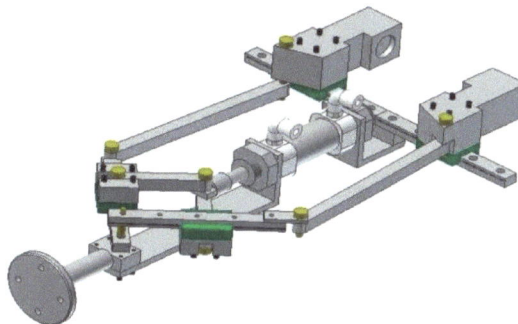

Figure 11: 3D-CAD model of the two-fingered pneumatic gripper of Fig. **4** [6].

(a)

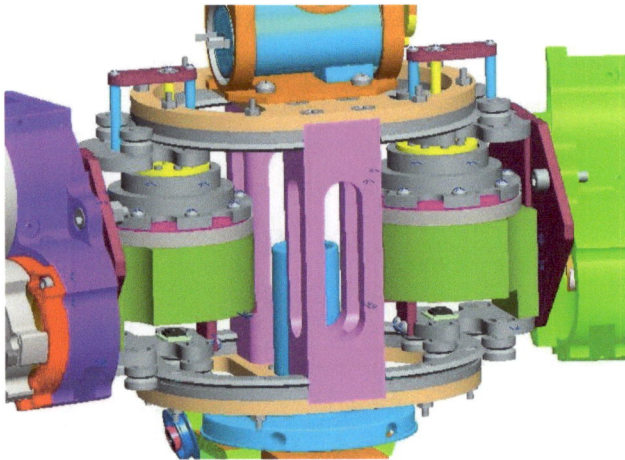

(b)

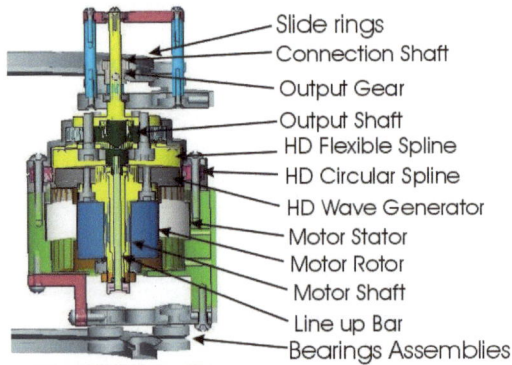

Slide rings
Connection Shaft
Output Gear
Output Shaft
HD Flexible Spline
HD Circular Spline
HD Wave Generator
Motor Stator
Motor Rotor
Motor Shaft
Line up Bar
Bearings Assemblies

(c)

Figure 12: Mechanical design of a new chest sub-system for the humanoid robot iCub [21]: (a) an overview of the upper-body including the proposed chest; (b) detail of the chest design; (c) cross-section of the actuating joint.

Fig. **12(a)** shows a proposed chest mechanism for the humanoid robot iCub [21]. This mechanism provides roll movements for the whole arm structures actuated by only one DOF for each sub-system. The mechanism is composed of two-90-deg segments of HepcoMotion PRT slide rings [22] installed in a chest frame. Two Kollmorgen RBE-01211 motors [23] assembled with two Harmonic Drive CSD-17-100-2A-GR gear boxes [24] actuate the chest joints generating roll movements for carrying the two arm sub-systems.

Both motors and Harmonic Drive have been selected as frameless version in order to design the joint housing with suitable shape for the chest assembly design. The Motor + Harmonic Drive assembly generates about 40 Nm of actuating torque for each chest joint. This actuating torque usually fulfils the carrying operation of an arm sub-system of a humanoid robot plus a medium-weight payload. Fig. **13(b)** shows a detail of the chest joint structure. The motor housing is connected to the slide rings through HepcoMotion bearing assemblies [22] that are installed in two bases and located one in the top and the other in the bottom of the joint. Fig. **13(c)** shows a cross-section of the chest joint in which main joint components are indicated. The compactness of the join design is possible to realize by checking Fig. **13(c)**. Note that a bar is used to keep the motor shaft and the output shaft lined up. This solution ensures a safe operation under high velocity conditions. SKF ball bearings [25] support the joint shafts and the line up this bar. Humanoid arm sub-systems are installed in the chest mechanism by two connection bases. The shape of these bases can be designed according to the specific humanoid arm structure in order to obtain an adequate chest-arm connection. The chest mechanism design is modular, which means, it can be installed and uninstalled form the robot structure depending of the humanoid project goal. The proposed chest sub-system shows a compact design with easy-operation and can be a successful solution in order to increase the workspace of the robot upper-body.

Taking into account the problems detected during experimental experiences with the current iCub head, a new sub-system has been designed in order to improve the robot mechanical structure and performance. Fig. **13** shows a 3D-CAD model of the new iCub head design, still divided in three parts: neck, eye and cover. In the new design, the cover has the same toy-like structure of the current head design. The neck sub-system is still configured in a serial 3 DOF manipulator with the joints located with the same order of the current design (pitch yaw and roll from bottom to top). The actuation of neck joints of Fig. **13(b)** is composed of Faulhaber DC motor with Harmonic Drive. The new neck design is robust, accurate and easy to operate as well as can perform human-like movements for the iCub head operation. Similar, Fig. **13(c)** shows the new design of the iCub eye

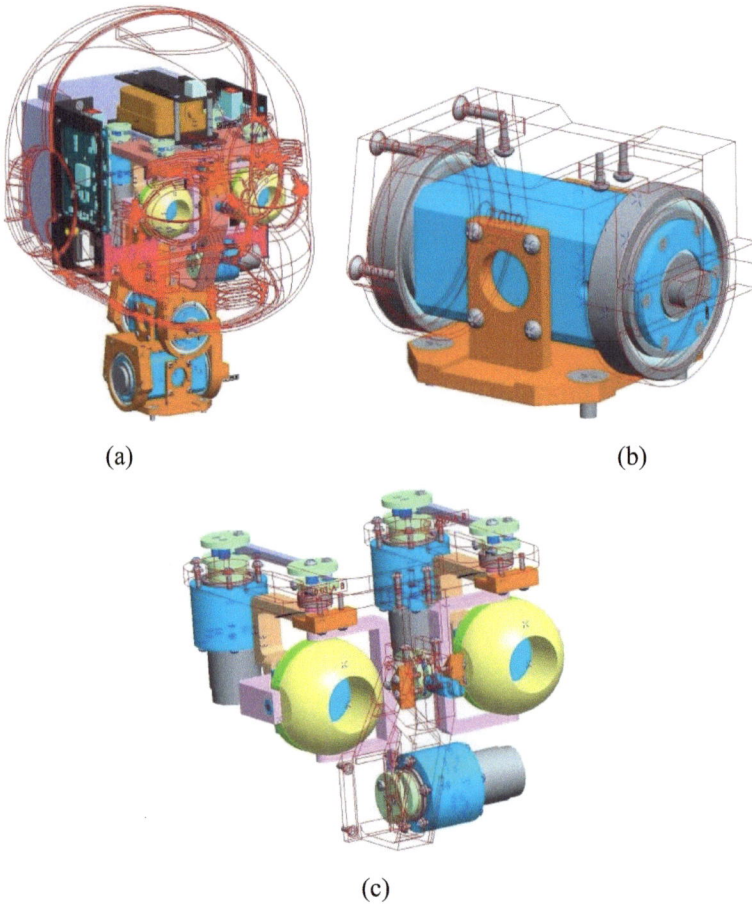

(a)

(b)

(c)

Figure 13: A new design for the head sub-system of the humanoid robot iCub [26]: (a) overview of the design; (b) a neck joint structure; (c) detail of the vision system.

sub-system. The new eye is a parallel manipulator of 2 DOF in which the tilt is still a common DOF and the pan is an independent DOF for each eye. The serial configuration of the current eye design has been changed to a parallel configuration since the parallel manipulator is high-speed and robust as well as provides proper human-like eye movement. The eye DOF are still actuated remotely by Faulhaber DC motors with Harmonic Drives in a similar configuration to the neck joint actuation. Moreover, the Harmonic Drives provide low-backlash and reduce the weight of the system. Four-bar linkage mechanisms transmit the motion from the motors to the eye moving platform, as shown in Fig. **13(c)**. Each eye has its own driving articulated mechanism since individual pan actuation is prevented. The articulated mechanisms are low-backlash, rigid and robust. Those characteristics are suitable for a vision system.

The mechanical design provides the shapes and materials of parts that compose the structure as well as their assembly including the commercial components. The actuation system is also selected during the mechanical design stage that provides more information about the dynamic characteristics of the robot.

COMPUTER ASSISTED VALIDATION

Since new characteristics for the robotic system, unconsidered during kinematic and dynamic study, have been introduced into de design process during the mechanical design, an operating validation of the obtained structure is necessary. For this aim, the use of dynamic simulation software is completely required for the development of virtual models in virtual environments. For example, a developed 3D model of the robotic finger of Fig. **3** has been used in order to check the feasibility of assembling all the components. This model has been used in order to carry out several simulations of the finger operation. These simulations have been useful in order to check the feasibility of the proposed dimensions for avoiding any link interference and for properly mimicking the human index finger motion. Similar, an ADAMS model of Fig. **11** gripper has been used in order to carry out a simulation of operation performance. The operating simulations have been considered for checking the functionality of the new gripper before its construction. The dimensions of the 3D-CAD model of Fig. **11**, mechanical characteristics, component materials, joints and actuation characteristics have been introduced for the ADAMS simulation in order to obtain the most approximate simulation. Two actions have been simulated, such as the gripper grasps the object without perturbation and the grasped object is hit by another object during the grasp, as shown in Fig. **14**, thus an impulsive action is generated. Fig. **15** shows the frame XYZ used as reference for obtaining the kinematic and dynamic parameters of the mechanism during the simulation. The gripper model in ADAMS environment of Fig. **14** takes into account several aspects, such as external forces, gravity, contact constraints, friction and inertia properties. Successful results of the dynamic simulations, which validate the model operation, have been reported in [6].

An ADAMS model of the final version of CALUMA robot of Fig. **9** has been used in order to carry out dynamic operating simulations during three basics operation modes, as reported in [8] and [27]. A first mode of dynamic simulation is related to the straight walking operation, as shown in Fig. **16**. The simulation has consisted in several steps of CALUMA for a human-like motion with a velocity of 0.5 sec/step on a flat terrain. The computed size of the CALUMA step is about 0.3 m. The leg module must be able to balance statically and dynamically the weight of the body by maintaining the zero movement point (ZMP) [28] within the support area.

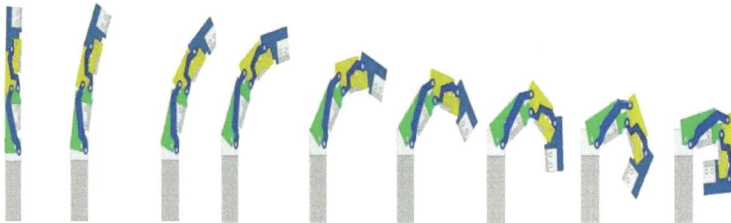

Figure 14: Simulation of the closing movement for the finger of Fig. **3** [5].

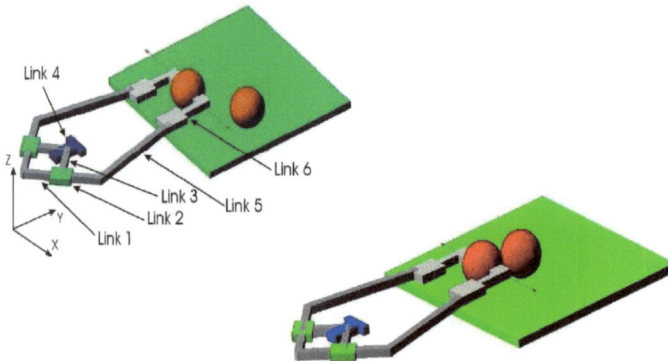

Figure 15: Dynamic simulations of the gripper of Fig. **12** operation [6].

Contribution of trunk and arm modules is also important in order to balance the robot structure during the walking operation. By coordinating properly the movements of legs, trunk and arm sub-systems, the robot can satisfy the condition of ZMP during walking. A method for generating the walking movement has been elaborated as based on replicating a human walking. It consists of synchronization among legs and arms, thus when the left arm and right leg move forward, the right arm and left leg move backward. A simulation of CALUMA operation while carrying an object has been elaborated as a second mode of dynamic simulation, as shown in Fig. **16(b)**. The walking simulations have consisted in several steps on a flat terrain with a velocity of 0.5 sec/step. The computed size of the CALUMA step is about 0.3 m. The grasped object has 1.5 kg weight, 30 mm of height, 60 of mm depth and 450 mm of width. This simulation has been developed by considering balance strategies that maintain the ZMP conditions too. During this CALUMA simulation the arm sub-systems do not move because they carry the grasped object, thus the trunk sub-system has to give a mayor contribution to the robot balance. The low number of CALUMA active DOFs for this simulation mode limits the balance capability of robot since the arm movement is necessary for the equilibrium of the structure, as above-mentioned. In fact, in order to carry an object during walking on a straight line with a proper velocity, the control on the CALUMA trunk sub-system should be modified when arms are holding the grasped object. Nevertheless, CALUMA presents a suitable performance for carrying a load during the walking according to the actuators capabilities, as shown in the results [27]. The third mode of dynamic simulation has consisted in a lift and release operation of CALUMA, as shown in Fig. **16(c)**. During this simulation, the manipulated object has been the same of the walking while carrying an object simulation. The grasping simulation has consisted in CALUMA approaching the object by using the trunk and arm sub-systems. Then, a firm grasp is obtained by using the hand sub-system. Finally, lifting and release of the object is achieved by using the trunk sub-system. The total operation time is 3.6 sec. The results of the dynamic simulation, reported in [8] and [27], validate the operation of CALUMA robot for straight walking as well as pick and place operations.

Fig. **17** shows a sequence of the simulation for the proposed iCub chest operation of Fig. **12**. The proposed chest model has been carefully developed taking into account the compactness and precision of the design that can be checked during dynamic simulation of the robot performances.

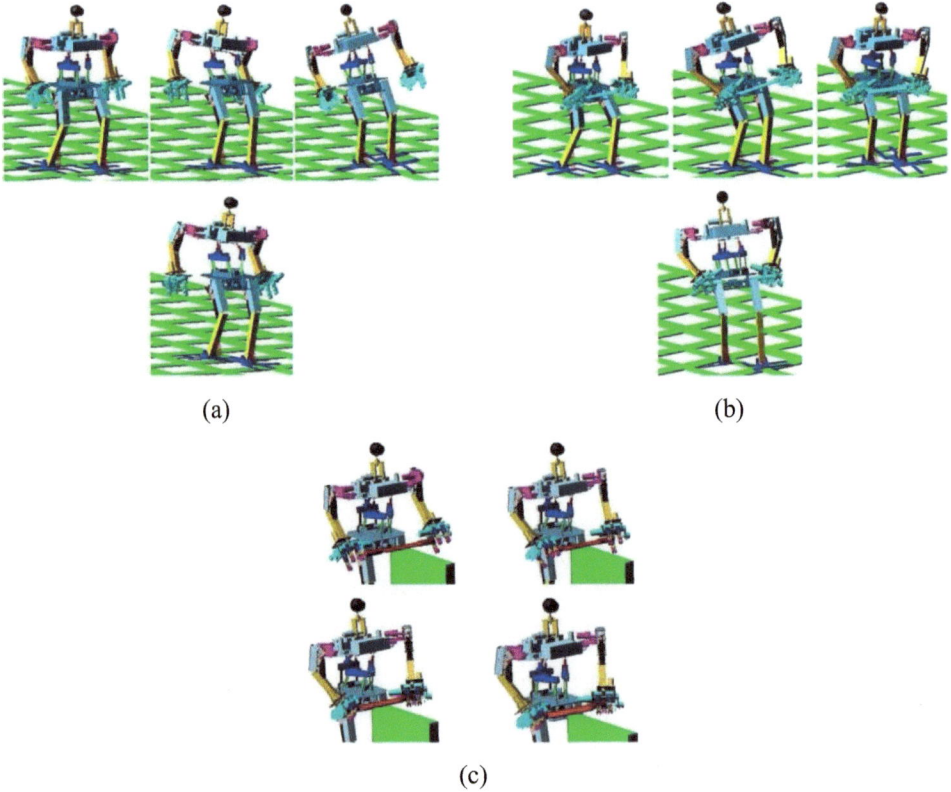

(a)   (b)

(c)

Figure 16: Dynamic simulation of three basic modes of CALUMA operation [27]: (a) walking; (b) walking with a grasped object; (c) grasping an object.

Figure 17: A performance sequence of the dynamic simulation for the proposed chest of Fig. **12** [21].

The dynamic simulations involve an iCub model in a chest operation that carries the arm sub-systems in a fully-open configuration. Several simulations of chest operations have been developed in which the arms acquire different configurations. Nevertheless, the worst design condition that requires maximum operating power is to bear the arms fully-opened, as it has been simulated. The results of the dynamic simulations have illustrated the successful performance of the proposed mechanism. Smooth curves, cyclic evolutions and suitable values have been common characteristics of the result plots, as reported in [21]. The possibility of a practical application of the proposed chest design in a built humanoid prototype has been validated by the simulation results.

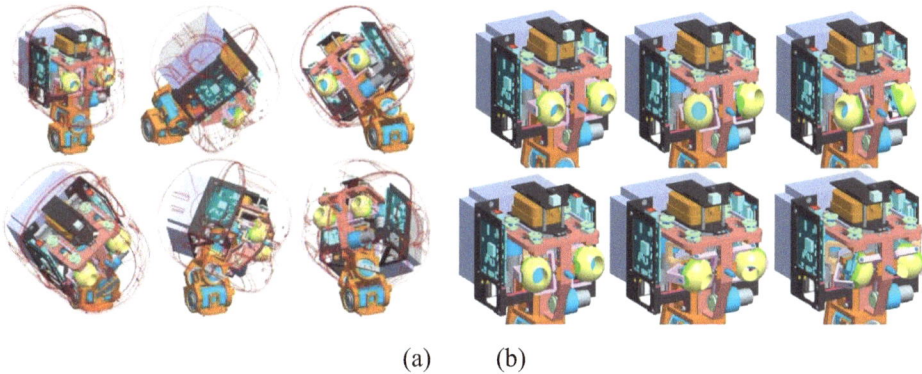

(a)   (b)

Figure 18: Dynamic simulations for the validation of the new proposed design of iCub head of Fig. **13** [26]: (a) neck sub-system; (b) vision system.

By using the 3D-CAD model of Fig. **13**, simulations of the new iCub head have been developed in Pro/Engineer environment in order to validate the feasibility of the assembly and check the operation under dynamic conditions. Mechanism Toolbox of Pro/Engineer [29] has been used for dynamic simulations due the convenient features for simulating the operation of multi-body systems that have been modelled in Pro/Engineer. The used mechanism Toolbox of Pro/Engineer computes backward differentially the algebraic equations of the dynamic model. The iCub head model in Pro-Engineer environment takes into account several aspects such as, external forces, gravity, contact constraints, friction and inertia properties. Friction effects are assumed negligible for the iCub joints in the simulation since low-friction bearings [25] have been used in the robot design. Almost all iCub head components have been modelled as made of aluminium alloy. Components, such as the cover and electronic boards, have been modelled by representing their material properties as accurately as possible. The dynamic simulations involve the simulated iCub in two modes. The neck and eye operations have been simulated separately since, as mentioned before, the model has been considered divided in neck, eye and cover. A first mode of dynamic simulation is related to a neck operation in which all 3-DOF are actuated, as shown in Fig. **18(a)**. The second mode of dynamic simulation is related to the eye operation, as shown in Fig. **18(b)**. The time interval of 0.01 sec (100 intervals in 1 sec) has been selected for the dynamic simulations in order to obtain suitable results within reasonable computational periods. The total simulation time was 2.00 sec. The results of the dynamic simulations of Fig. **18**, [26], validate the proper characteristics of the new head design as sub-system of the iCub robot.

When the results of dynamic simulations validate a virtual robot operation, the current design can be assumed as final. Therefore, the prototyping process can be carried out consequently.

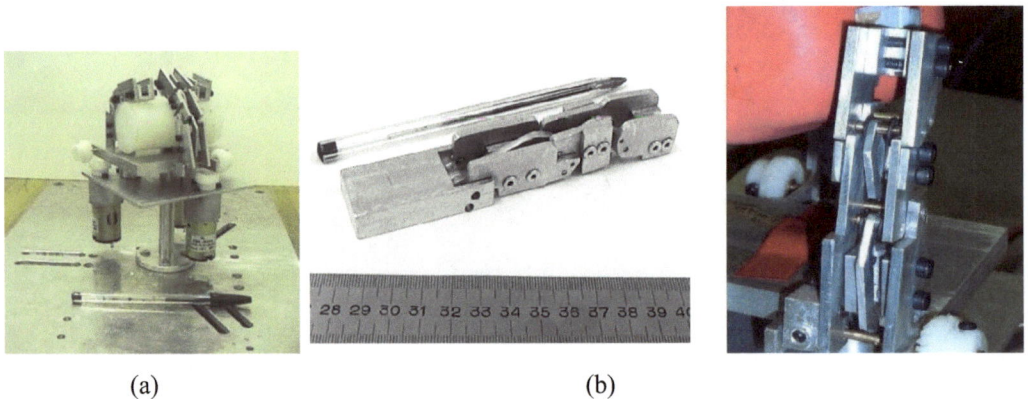

(a)   (b)

Figure 19: The three-fingered anthropomorphic hand LARM Hand of Fig. **10(a)** [5]: (a) an overview of the mechanical structure of whole hand; (b) details of a finger structure.

PROTOTYPING

Fig. **19(a)** shows a built prototype of the robotic hand of Fig. **10(a)**. This robotic hand has been constructed by considering the concept of low-cost and easy-operation. It is composed mainly of commercial components that can be found in the market. Fig. **19(b)** shows details of the finger structures. Note the articulated mechanism of Fig. **3** inside the body of the finger that provides robustness and compactness. This robotic hand presents approximately the dimensions of an average human hand and fingers have the dimensions of an average human index finger. Fig. **20** shows a prototype of the two fingers gripper of Fig. **11**. Similar to the hand prototype of Fig. **19**, this gripper prototype is mainly composed of commercial parts.

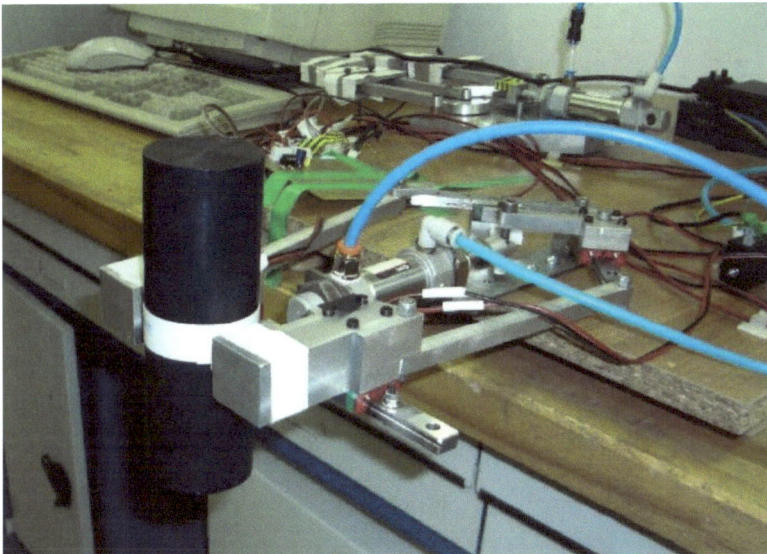

Figure 20: A built prototype of the pneumatic gripper with two fingers of Fig. **11** [6].

The prototype of Fig. **20** contains the pneumatic system for the actuator operation. Moreover, two force sensors have been installed on the finger tips in order to manager the grasp force preventing a close-loop control strategy.

In 1998, a multi-disciplinary initiative of collaborative research focussed on artificial cognitive systems was started by the European Commission. One of the results of this initiative was the humanoid robot project RobotCub [30]. ICub is a humanoid robot that has been created as an open and freely-available humanoid platform for research in embodied cognition. ICub has the size and shape similar to that of a three-and-a-half year-old child, as shown in Fig. **21**. One of the aims of RobotCub is to achieve cognitive capabilities through developmental and learning processes, such as interactive exploration of its environment, manipulation, imitation and gestural communication [30].

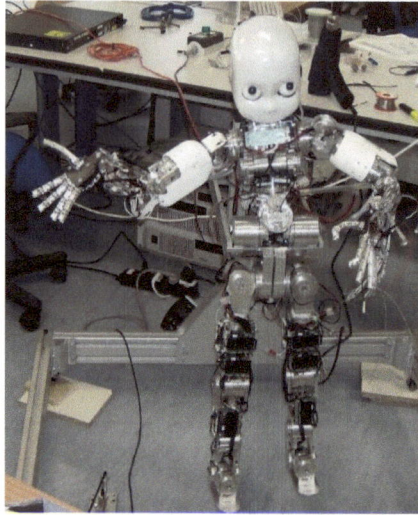

Figure 21: The humanoid robot iCub [30].

EXPERIMENTAL VALIDATION

Experimental tests with the hand prototype of Fig. **19** have been developed during tip and cylindrical grasping of certain objects. Fig. **22(a)** shows the tip-grasp experience of the LARM hand grasping a compliant plastic object. During tip-grasp experiences sensors have been installed on the finger tips in order to monitor the forces during experiments. In the result plots of tip-grasp experiences, time evolution of the grasp forces does not show stable stage since the contact between grasped object and sensors has not been completely firm. Nevertheless, the plots of forces have presented suitable results for the tip-grasp of LARM Hand since the grasp has shown firm configuration with proper level of forces and phase durations, as reported in [31]. Fig. **22(b)** shows the LARM hand during the cylindrical grasping test. The phase operations during cylindrical-grasp experiences have been the same of the tip-grasp. Force sensors have been installed in every phalanx and palm surfaces. The operation of the sensors and the DC motors in this prototype has been managed with an acquisition card PCI 6024 E and a suitable virtual instrument developed in LabView by following the scheme of Fig. **23(a)**. Both in the results of cylindrical-gasp and tip-grasp tests, the plots show proper magnitudes of measured forces and time evolution. The time evolution of the grasp forces has presented a similar evolution of increase, remaining and decrease in both grasp types [31] and [32]. This validates the effectiveness of the LARM hand as a grasping robotic device.

The gripper prototype of Fig. **20** has also been validated through experimental tests of different objects grasping. According to the scheme in Fig. **23(b)**, an acquisition card, which is operated by a virtual instrument, measures force signal from finger sensors. A commercial PLC manages the closed-loop force control block of Fig. **23(b)** that has been used for regulating a prescribed gripping force value. In the control scheme of Fig. **23(b)**, the signal of a force sensor F_{out} is used as feedback analogue signal as well as for monitoring the action of the grasping force. A PI (Proportional-Integral) control has been used. The proportional integrative control scheme has been chosen to suitably regulate the grasping force at a stable determined level. F_{ref} is the prescribed force and it is included in the control algorithm. F_{ref} can be previously calculated experimentally in order to grasp and manipulate the objects without damages.

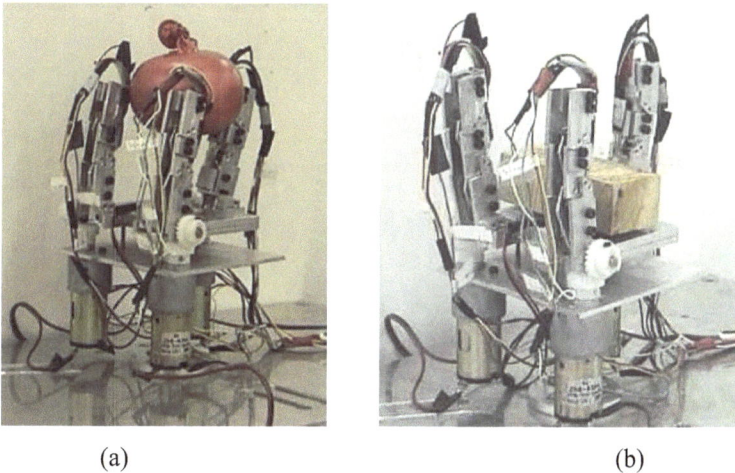

(a)   (b)

Figure 22: Experimental validation of LARM Hand: (a) during tip grasping [31]; (b) during cylindrical grasp [32].

The control-input variable e is the error signal that is given as the difference between the desired force F_{ref} and the current value F_{out}. The force correction has two terms, a straight proportional signal and a scaled time integral of the error. By developing iterative proves the most convenient magnitudes of the integral

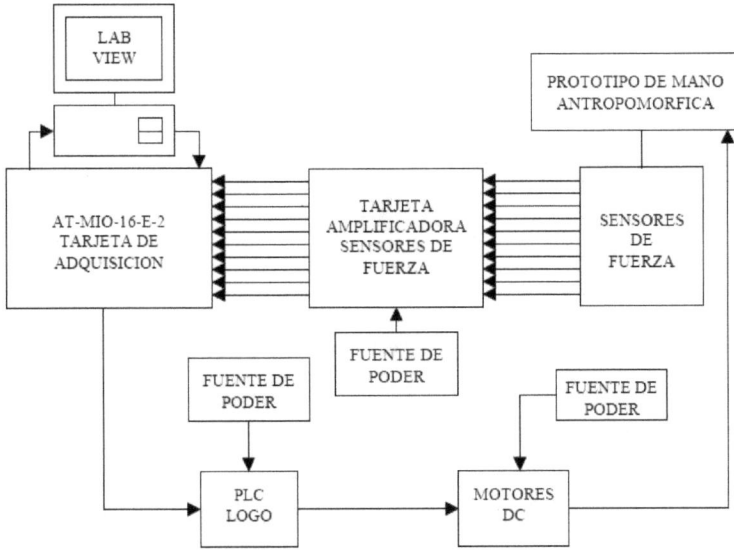

(a)

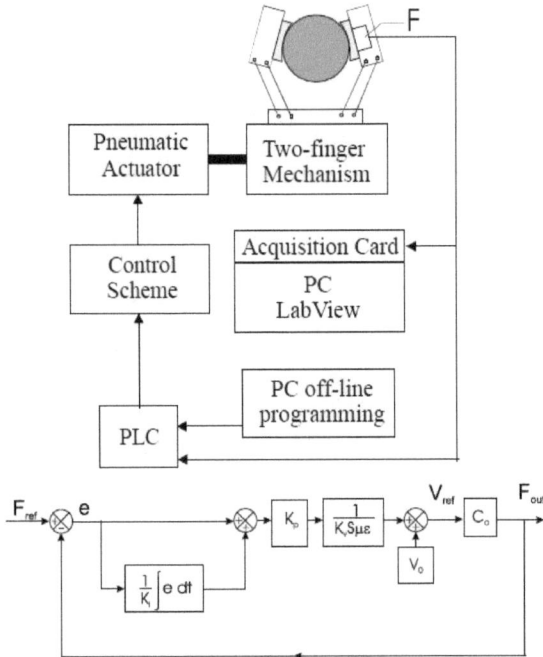

(b)

Figure 23: Mechatronic system used for the experimental validations of the grasping devices of Figs. **19** and **20**: (a) a scheme for the hand validation of Fig. **21** [31]; (b) a scheme for the Fig. **20** gripper [6].

coefficient K_i and proportional coefficient K_p can be evaluated. By using suitable magnitudes of K_p and K_i, the error signal is rapidly approaches to zero. Indeed, the integral action reduces the error to zero depending of the setting of the gain K_i and K_p. In Fig. **2** K_v, V_0 and C_a are respectively the static characteristic, dead band and amplification coefficient of the pressure proportional valve; S is the area of the piston surface; μ is the correction factor due to friction effects; and ε is the efficient coefficient of the driving mechanism that is expressed in percent. Therefore, the mechatronic approach have been recognized when mechanical efficiency, friction and size coefficients have been used together in the control gains to evaluate the force regulation successfully, as reported in [6].

The humanoid robot iCub of Fig. **21** has been used in the recent past in order to develop cognitive strategies for carrying out different tasks, as above-mentioned. During these activities, some problems in the robot structure have been detected since they make more difficult the successful task computing [26]. Although the current neck design is robust, easy to operate and highly performing, it presents the following problems:

- The motors barely fit the torque requirements for the head movement operation. A torque quite bigger than the just required for a free robot movement is necessary in some control applications;

- Neck joints contain bushes, as shown in Fig. **24(a)**, that generate high friction, leading to loss of mechanical energy;

- The high reduction ratios of the current planetary gearboxes results in low-velocities, as shown in Table **1**, that do not fulfill the robot application requirements;

- The current planetary gearboxes present some backlash that penalizes the robot accuracy.

The eye mechanism has a serial configuration with 3 DOF. Both eyes can pan independently by using 1 DOF each one and they can tilt simultaneously by using a common DOF. Fig. **24(b)** shows in detail the eye mechanism of the current iCub head. Each DOF of the eye is actuated remotely by a Faulhaber DC motor with planetary gearbox through a belt mechanism. The pan movement is driven by a belt mechanism with the motor located at the eye ball. The tilt is actuated by a belt mechanism located between the eyes. Each belt has a tension adjustment mechanism. For the necéssary acceleration and speed, the DC motors equipped with optical encoders have comfortably larger specifications than required. Nevertheless, the current iCub eye sub-system presents the following design problems:

- The belt system presents slippage and backlash since the force is transmitted by the contact between belt teeth;

- In highly dynamic applications, the belt system can generate some vibrations, unsuitable for the vision system, which requires frequent adjusted of tension;

- The fact that the tilt motor has to carry the pan motors slows the system down;

- A small amount of backlash is always necessary to reduce excessive wear, heat and noise created by the current gearboxes.

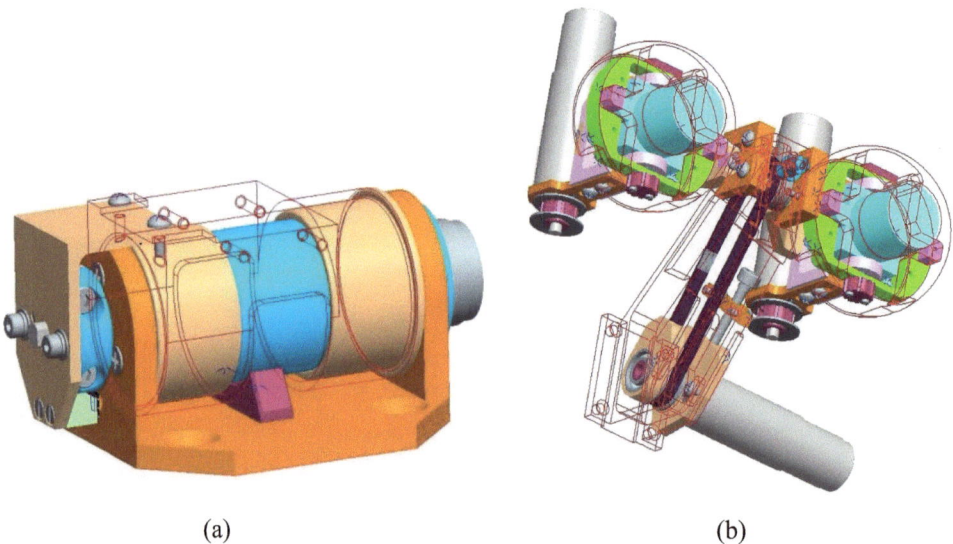

(a)   (b)

Figure 24: Sub-assemblies of the current iCub head design [12]: (a) a neck joint; (b) vision system.

Therefore, a new design process have began from the mechanical design of a new head sub-system by using the information of the problems found out during experiences with the built prototype of Fig. **21**. The result of this new iteration has been the head model of Fig. **13** that have been validated thought the dynamic simulations of Fig. **18** [26]. A future work will be including a built prototype of the new iCub head in the whole robot structure.

OPTIMIZATION PROCESS

Optimization is a process composed of a mathematic algorithm providing solutions that can mimic an optimized model. It is worthy to emphasize that optimization can be confused with improvement when some topics of the conceptual design is enhanced. The optimization can be elaborated after obtaining the final result of a design process starting a new design. Nevertheless, the proposed design procedure of Fig. **1** can begin from an optimization of a certain robot system. For example, an optimization process has been applied on the robotic finger of Figs. **3**, **15** and **20(b)**, reported in [12]. The design of an anthropomorphic finger must fulfil basically requirements of:

- Human-like motion and grasp;

- Compact size;

- Actuation lightweight and efficiency;

- Position and force control.

The above-mentioned aspects can be evaluated numerically for analysis and design purposes. In particular, human-like motion can be described by looking at the deviation of the motion characteristics between a human test and an operation of a designed finger mechanism. Following the above-mentioned design requirements and criteria, the design of finger driving mechanisms can be conveniently attached by considering them in a simultaneous way. An optimum design procedure for 1-D.O.F. finger driving mechanisms can be formulated by using the above-mentioned evaluation criteria for finger design in a multi-objective optimization problem in the form

$$\text{min } \mathbf{f(X)} \tag{4}$$

subject to

$$\mathbf{G(X)} \geq 0 \tag{5}$$

in which $\mathbf{f}$ is a vector function whose components are the design optimality criteria; $\mathbf{X}$ is the vector of design parameters; and $\mathbf{G}$ is a vector function of constraints due to the application and characteristics of the design problem. The numerical solution of a multi-objective optimization problem can be very complex and computationally troublesome. Several techniques are available, [33], even in commercial software packages. The solution with weighting factors can be considered suitable from engineering viewpoint, also because it permits the designer to guide and track the significance of the optimality criteria by adopting numerical values of the weighting factors in agreement with his/her engineering experience and problem peculiarities. The above-mentioned optimality criteria can be used as objective functions in the form of normalized computations and therefore the optimum design of driving mechanisms for 1-D.O.F. anthropomorphic fingers can be formulated specifically with a weighted objective function through the weight coefficients w_i (i=1,2,3) in the form

$$\text{min } \mathbf{f(X)} = w_1 \, \mathbf{f_1(X)} + w_2 \, \mathbf{f_2(X)} + w_3 \, \mathbf{f_3(X)} \tag{6}$$

subject to

$g_1 = l_{ji} > 0 \ (i,j=1,2,3);$

$g_2 = l_{4i} < d_h \, (i=1,2);$

$g_3 = l_{12} < d_h;$

$g_4 = l_{12} > |l_{51} - l_{31}|;$ (7)

$g_5 = l_{41} > |l_{52} - l_{32}|;$

$g_6 = l_{21} > (l_{11}^2 + l_{51}^2 + l_{12}^2 + l_{41}^2)^{1/2};$

$g_7 = l_{21} < (l_{11}^2 + l_{51}^2 + l_{12}^2 + l_{41}^2)^{1/2};$

$g_8 = l_{22} > (l_{12}{}^2 + l_{52}{}^2 + l_{42}{}^2 + l_{41}{}^2)^{1/2}$;

$g_9 = l_{22} < (l_{12}{}^2 + l_{52}{}^2 + l_{42}{}^2 + l_{41}{}^2)^{1/2}$.

with

$$f_1 = \Delta\varphi_2 + \Delta\varphi_3 ;$$

$$f_2 = \frac{\Delta L}{L_h} ;$$ (8)

$$f_3 = \frac{\Delta\tau}{\sum\limits_{i=1}^{n} \tau_{hi}} .$$

In particular, f_1 in Eq.(6) represents the optimal criterion for finger human-like motion, in witch $\Delta\varphi_i$ (i=1,2) is the sum of $h_{\theta i}$ and $f_{\theta i}$ (i = 2,3) that are obtained in Eqs. (1) and (2). The expression f_2 in Eq.(6) represents the compact sizes optimal criterion, in which ΔL is the size difference between the human finger and mechanical finger. Finally, f_3 describes optimal criterion, for static equilibrium and energy consumption in which $\Delta\tau$ is the torque difference between the actuating torque for the driving finger mechanism and actuating torque for the human finger. The non-dimensional expressions are useful to consider the kinematic design of the finger mechanism independently of the overall sizes. However, the specific dimensions are imposed by the constraint expressions in Eqs. (7). Equations (7) represent the design constraints for the multi-objective optimization problem in an optimum design procedure for 1-D.O.F. finger driving mechanisms. Particularly, the first constraint equation is positive definite in order to ensure link parameters of the driving mechanism with positive definite practical values. The second and third equations represent human finger size constraints for the link parameters when d_h is given as thickness of the finger body. The fourth to ninth equations are geometric constraints for the link parameters that are necessary to build the finger driving mechanism according to the scheme of Fig. **2**.

The kinematics of the motion of human fingers can be analysed by using video recording and photo sequences. Then, the observed motion can be described by

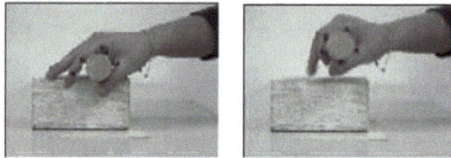

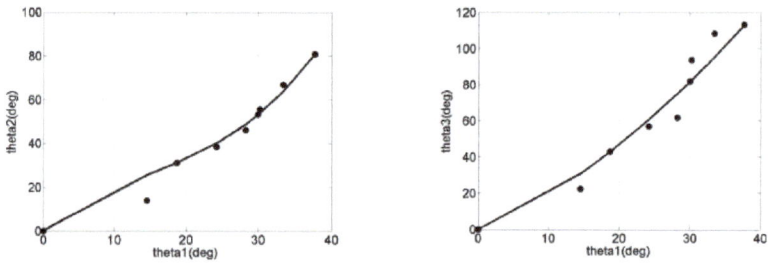

(a)

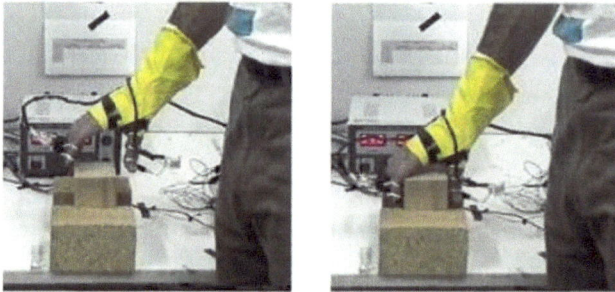

(b)

Figure 25: Study of the human grasp for the optimization process of Fig. **26**: (a) a cylindrical grasp performance [12]; (b) measurement of cylindrical grasp forces [5].

looking at the trajectory of reference points and time function of the articulation rotations. Indeed, the trajectory can be represented as a path function of the articulation joints only. Fig. **25(a)** shows a photo sequence of a human finger motion in approaching the grasp of a cylinder.

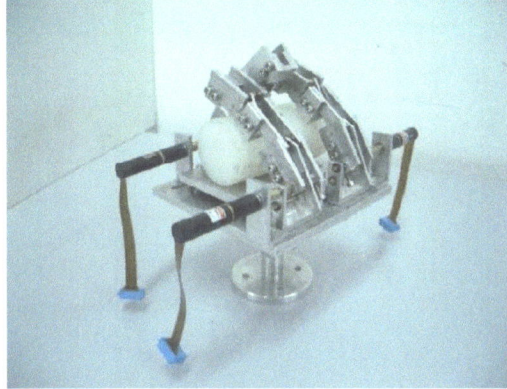

Figure 26: The robotic hand as results of the optimization process.

The evolution of the motion has been obtained by looking at the motions of marks on the articulations and phalanx links that have been glued on the finger surface. In the example for design purposes, nine different configurations have been measured experimentally, as shown in Fig. **25(a)**. The continuous line is a cubic interpolation of the experimental data that are reported as circular black dots. Similar results have been obtained for the grasp of different objects, and they are reported in [5]. A numerical evaluation of the human-like motion can be obtained by using an anthropomorphic finger mechanism. The numerical evaluation can be obtained by looking at the joint motion as compared to a finger design scheme. The experimental data in Fig. **25(a)** can be used to identify an interpolating cubic line for the phalanx joint angles θ_2 and θ_3 as functions of θ_1 joint angle of the first phalanx that can be considered as the inputted 1-D.O.F. motion. Thus, an interpolating cubic line can be expressed as

$$h_{\theta i} = c_{1hi}\theta_{1h} + c_{2hi}\theta_{1h}^2 + c_{3hi}\theta_{1h}^3 \qquad (9)$$

in which the coefficients c_{jhi} (j=1,2,3; i=2,3) are determined by using the experimental data.

Fig. **25(b)** shows grasp experience of different objects by human hands. These tests have been carried out by different persons following a fixed procedure. The sizes criterion f_2 and the criterion for static equilibrium and energy consumption f_3 have been evaluated through Eq. 8 by measuring the sizes of the persons involved in the experience of Fig. **25(b)** and their grasping forces, respectively. The dynamic scheme of Fig. **7(a)** has been considered for the formulation of the f_3 optimal criterion. Fig. **26** shows an optimized robot hand that has been the result of this process.

CONCLUSIONS

A novel procedure for the mechanical design of robotic systems has been presented. This procedure is composed of certain phases in which important topics of the structure design are studied. The proposed strategy has been decrypted by using illustrative examples of robots developed in the latest years, such as a robotic hand with 1-DOF fingers, a pneumatic gripper with two fingers, a low-cost humanoid robot and iCub robot. The evolution of these designs has been recognized at the end of every phase. This approach can be an iterative process in which some problems, identified during validations, should be resolved during the proper stage of the conceptual design given the beginning of a new design process, as occurred for the iCub robot design. An optimization has been prevented, as part of the proposed approach, which provides higher quality for designed systems from the mechanical point of view. The novel strategy is no-complex and effectively gives suitable solutions that can be applied in a practical application of any robotic system design.

ACKNOWLEDGEMENT

Declared none.

CONFLICT OF INTEREST

The author(s) confirm that this chapter content has no conflict of interest.

REFERENCES

[1]   P. Childs, *Mechanical Design*. Elsevier Butterwoth-Heinemann: 2nd Edition, 2004.

[2]   M. Ceccarelli, "Low-Cost Robots for Research and Teaching Activity", *IEEE Robot. Autom. Mag.,* vol. 10, no. 3, pp. 37-45, 2003.

[3]   N.E. Nava Rodriguez, E. Ottaviano and M. Ceccarelli, "Workspace Analysis of RRP Manipulators", In: *European Conference on Mechanism Science*, Obergurl, 2006, paper EuCoMeS-035.

[4]   N.E. Nava Rodriguez and M. Ceccarelli, "Región Posible para el Espacio de Trabajo de un Manipulador Genérico de Tres Pares de Revolución", *Iberoamerican J. Mech. Eng.*, vol. 11, no. 2, pp. 67-80, 2007.

[5]   M. Ceccarelli, N.E. Nava Rodriguez and G. Carbone, "Design and Tests of a Three Finger Hand with 1 Dof Articulated Fingers", *Int. J. Robot.*, vol. 24, pp. 183-196, 2006.

[6]   N.E. Nava Rodriguez, C. Lanni and M. Ceccarelli, "Design, Simulation and Validation of a New Two-Fingered Gripper with Pneumatic Actuator", *Int. J. Robot. Manage.,* vol. 11, no. 1, pp. 9-14, 2006.

[7]   A. Erdman and G. Sandor, *Mechanism Design: Analysis and Synthesis*. Prentice-Hall: New Jersey, 1984.

[8]   N.E. Nava Rodriguez, *"Design and Simulation of a New Low-Cost Easy-Operation Humanoid Robot"*, PhD Thesis, University of Cassino, Italy, 2007.

[9]   E. Ottaviano, C. Lanni and M. Ceccarelli, "Numerical and Experimental Analysis of a Pantograph-Leg with a Fully-Rotative Actuating Mechanism", In: *World Congress in Mechanism and Machine Science*, Tianjin, 2004, pp.1537-1541.

[10]   *Wikipedia homepage.* Available: http://en.wikipedia.org/wiki/Golden_ratio [Accessed: 2006].

[11]   L. Tsai, *Robot Analysis*. John Wiley & Sons: New York, 1999.

[12]   N. E. Nava Rodriguez, G. Carbone and M. Ceccarelli, "Optimal Design of Driving Mechanism in a 1-d.o.f. Anthropomorphic Finger", *Mech. Mach. Theory*, vol. 41, no. 8, pp. 897-911, 2006.

[13]   M. Ceccarelli, *Fundamentals of Mechanics of Robotic Manipulation*. Kluwer Academic Publishers Group: 2004.

[14]   MSC.ADAMS, *"Documentation and Help"*, User CD-ROM, 2005.

[15]   N.E. Nava Rodriguez, G. Carbone and M. Ceccarelli, "Design Evolution of Low-cost Humanoid Robot CALUMA", In: *World Congress in Mechanism and Machine Science*, Besancon, 2007, paper ID: 181.

[16]   *LARM homepage.* Available: http://webuser.unicas.it/weblarm/ [Accessed: 2010].

[17]   A. Maceli, *Scienza delle Costruzioni*. Accademica Roma: Vol. 1, 2nd Edition , 1999. (In Italian)

[18]   J. Shigley, C. Mischke and Budynas R., *Progetto e Costruzione di Macchine*. MacGraw-Hill: 2004. (in Italian)

[19]   D. Hanlon and W. Rainforth, "Some Observations on Cyclic Deformation Structures in the High-Strength Commercial Aluminium Alloy AA 7150", *Metall. Mater. Trans.,* vol. 29A, no. 11, pp. 2727-2736, 1998.

[20]   *Autodesk homepage.* Available: http://www.usa.autodesk.com [Accessed: 2010].

[21]   N.E. Nava Rodríguez, M. Abderrahim and L. Moreno, "A Mechanism Design of a Chest Sub-system For Humanoid Robot", In: *ASME International Conference of Advance Intelligent Mechatronics AIM2009*, Singapore, 2009, paper ID 43.

[22]   HepcoMotion PRT catalog, *Rings Slides and Track System*. 2008.

[23]   *Kollmorgen RBE(H) Series Motor Catalog.* Available: www.DanaherMotion.com [Accessed: 2010].

[24]   *HD homepage.* Available: www.harmonicdrive.com [Accessed: 2010].

[25]   *SKF homepage.* Available: www.skf.com [Accessed 2010].

[26]   N.E. Nava Rodriguez, "Design Issue of a New iCub Head Sub-system", *Robot. Comput.-Integr. Manuf.,* vol. 26, no. 2, pp. 119-129, 2010.

[27]   N.E. Nava Rodriguez, G. Carbone and M. Ceccarelli, "Simulation Results for Design and Operation of CALUMA, a New Low-cost Humanoid Robot", *Int. J. Robot.,* vol. 26, pp. 601-618, 2008.

[28]   M. Vukobratovic, V. Potkojak and S. Tzafestas, "Human and Humanoid Dynamic", *J. Intell. Robot. Syst.,* vol. 41, pp. 65-84, 2004.

[29]   S.G. Smith, *Pro/ENGINEER Wildfire 3.0 Update Manual for Wildfire 2.0 Users.* 2008.

[30]   N.E. Nava Rodriguez, V. Tikhanoff, G. Metta and G. Sandini, "Kinematic and Dynamic Simulations for the Design of ICub Upper-Body Structure", In: *9th Biennial ASME Conference in Engineering System Design and Analysis ESDA 2008*, Haifa-Israel, 2008, paper ESDA2008-59082.

[31]   N.E. Nava Rodriguez and M. Ceccarelli, "Tip-Grasp Experiences with Three-Finger Anthropomorphic LARM Hand", In: *International Workshop on Robotics in Alpe-Adria-Danube Region*, Balatonfured, 2006, paper BFD-006.

[32]   N.E. Nava Rodriguez, G. Carbone, E. Ottaviano and M. Ceccarelli, "An Experimental Validation of a Three - Fingered Hand With 1 Dof Anthropomorphic Fingers", In: *Intelligent Manipulation and Grasping*, Genova, 2004, pp. 285-290.

[33]   G.N. Vanderplaats, *Numerical Optimization Techniques for Engineering Design.* McGraw-Hill Publication: New York, 1984.

Send Orders for Reprints to reprints@benthamscience.net

CHAPTER 2

Automation in Order Picking System

Marco Melacini[*]

Department of Management, Economics and Industrial Engineering, Politecnico di Milano, Milan, Italy

Abstract: The new emerging trends in supply chain management have changed across the supply chain the role of the warehouse and the related activities. Warehouse managers report a significant increase in the number of SKUs (stock keeping unit) to manage, along with a reduction in size of orders delivered and, as a consequence, an increase in the incidence of order picking. Order Picking can be defined as the selective retrieval of unit loads from high level unit loads or of pieces or cases from racks in order to fulfill customer orders. Examples are the retrieval of cases from pallet unit load or the retrieval of pieces from cases. In this context growing attention is placed on questions such as "why automate?" and "where is the return?". Even though the automation in material handling systems began in the early '60s and '70, nowadays it is no more a question of fashion or innovation: deciding whether to automate or not usually requires a deep investigation on which factors need to be considered before automating. The topic of automation is even more critical with respect to order picking, where the importance and the complexity of the of the picking activity have induced material handling system suppliers to introduce increasing numbers of solutions with ever greater levels of automation. This chapter presents the main solutions to design Order Picking Systems (OPSs). In particular, after illustrating a general framework to design OPS, the pros and cons of the main automatic solutions for OPS are illustrated. The chapter concludes with the presentation of a taxonomy to support the choice of the OPS.

Keywords: Automation, picking, design framework, trends, classification, picking wave, overlapping effect, picker to parts, pick to box, pick and sort, parts to picker.

1. INTRODUCTION

A warehouse consists of three main areas: a receiving area, a shipping area, a storage area. In turn, the storage area can be subdivided in a reserve area, *i.e.* an area for storing the greatest part of the goods inventory, and a picking area (called also forward area), *i.e.* a smaller area designed with the aim to improve order picking

Address correspondence to Marco Melacini: Department of Management, Economics and Industrial Engineering, Politecnico di Milano, Milan, Italy; Tel: +39-02-2399.4059; Fax: +39-02-2399.4067; E-mail: marco.melacini@polimi.it

productivity and efficiency. Typically, in the reserve area full pallet handling is performed (*e.g.* retrieval of full pallets, replenishment of picking locations), while the forward area is characterized by broken and full-case picking. Picking activity is achieving an increasing key role in supply chain management, both from the point of view of the physical distribution activities and from point of view of production systems (*e.g.* the supply of assembly stations with kits). The picking activity involves significant costs and affects customer satisfaction levels. The service level is composed of a variety of factors such as the order delivery time, the order integrity and the order accuracy. In many cases, costs related to the order picking activity account for more than half of the total warehousing cost [1]. The importance and the complexity of the topic have induced material handling system suppliers to introduce increasing numbers of Order Picking Systems (OPSs) with ever greater levels of automation: "picker-to-parts", "pick-to-box", "pick-and-sort", "parts-to-picker", "completely automated picking" (*e.g.* robots or dispensers) systems. This greater automation has led to an increase in design complexity and to the risk of disruption and service level failings [2].

The complexity is due not only to the automation level, but also to the combination of other factors such as: product features (*e.g.* number, size, value, type of packaging, inventory level, sales), customer order profile (*e.g.* size, number of order lines, average daily number of orders) and operating policies for each functional area (*e.g.* pick by order or pick by item). This complexity is even more important at the design conceptualisation stage, in which designers explore alternative configurations and material handling technologies, and assess the performance of each solution. Finally, within the overall design process the identification of the most suitable OPS is one of the most critical phases, since it requires solving the trade-off between the completeness and the cost of the analysis [3].

2. ORDER PICKING SYSTEM DESIGN

As shown in Fig. **1** OPS design methodology can be composed by four stages:

- Input stage.

- Selection stage.

- Evaluation stage.

- Detail stage.

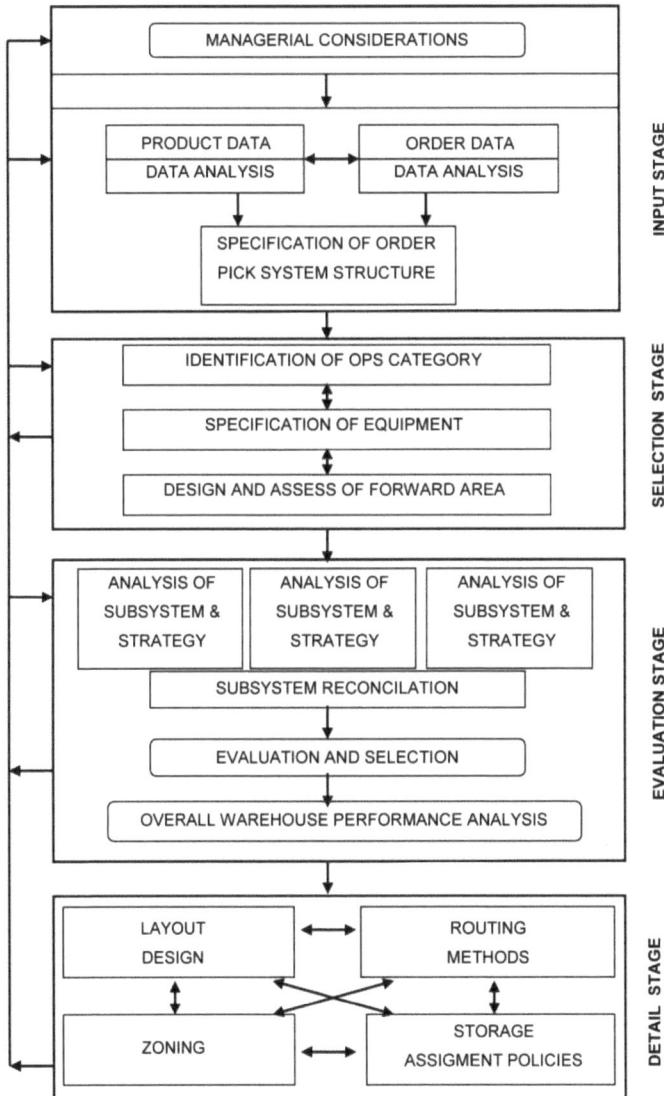

Figure 1: Methodology to design Order Picking System.

In the input stage are taken into account a few managerial considerations (*e.g.* budget, useful economic life, required payback time), operational constraints (*e.g.*

area available, ceiling height, number of daily shifts) and transaction data on customer order (*e.g.* daily number of orders, number of order lines for each order) and product features (*e.g.* number of items, ABC curve). The outcomes of the first stage are the specification of overall OPS structure (*i.e.* number of subsystems) and the definition of requirements for each subsystem (*e.g.* number of daily suborders, cases and order lines, number of SKUs). The selection stage consists of three sequential steps: OPS identification, specification of equipment types and design and assessment of the forward area. These three steps must be carried out for each identified subsystem. As previously noted , OPS identification is the topmost critical phase, often performed on the basis of insights and experience of warehouse designers. In the last section this topic will be further deeply investigated. Once the more suitable OPS has been identified, it is possible to compare different equipment types, in order to find the technical solution that best fits the problem. For example, once a "parts-to -picker" system has been chosen, it will be possible to evaluate the use of a horizontal carousel or a miniload, by defining the basic characteristics for each solution (*e.g.* if a single or double shuttle miniload should be employed).

Afterwards these two steps (OPS and the equipment type identification), it is necessary to design each subsystem, defining the number and size of picking locations and the size of the forward area. Once we know the size of forward area a first assessment of subsystem's performances (*e.g.* picking rate, response time) will be possible. The design of the picking area requires the resolution of the Forward Reserve Problem (FRP), deciding the quantity of each item to be placed in the forward area. Since the retrieval cost from the reserve area is much higher than the retrieval cost from the forward area, the overall picking cost can be reduced by assigning items to the forward area. On the other hand picking cost increases as the size of the forward area grows, while the replenishment cost decreases. Therefore the resolution of the FRP requires analysing the trade-off between picking cost and replenishment cost [4]. In the third stage of design methodology, the evaluation stage, the quantitative and qualitative reconciliation of the considered subsystems is performed. This activity may lead to a further selection and specification. Finally in the detail stage the characteristics of each subsystem are detailed, trying to optimise the performances as much as possible.

This requires an in-depth study of the aspects connected with layout design, storage and routing policies and zoning (*e.g.* zone size) in a context where the main characteristics of the system have already been defined. In order to maximise the outcome of the detail design it is advisable to consider the joint adoption of the aforementioned levers, due to the strict interdependency of their impact on performances. From an operational point of view, warehouse designers can use different approaches such as analytical models and simulation. During this stage some decisions previously made for the performance assessment may be refined (*e.g.* number of picking zones, batch size and length of pick-wave,) and the more operational issues of the OPS (*e.g.* routing policies, replenishment policies) can be treated in more details. These issues have also been considered in the previous phases, but with a lower level of detail. At this stage the issues deriving from the previous stages can be further studied and refined starting from the viewpoint of system performance improvement. The result of this optimisation process could lead to the outlining of new needs concerning the overall system framework. In this less likely case it will be necessary to go back and review, partially or wholly, the decisions made.

3. ORDER PICKING SYSTEM CLASSIFICATION

As previously mentioned, a variety of OPSs can be implemented in the forward area. In order to better understand the main features of each one, OPSs can be classified accordingly with four main decisions (Fig. **2**): who picks the goods (humans/machines), who moves in the picking area (pickers/goods), if conveyors are used to connect each picking zone and finally, which picking policy is employed (pick by order or by item). According to these criteria we can identify five main OPSs: "Picker-to-parts" system. "Pick-to-box" system. "Pick-and-sort" system. "Parts-to-picker" system. "Completely automated picking" system (*e.g.* robots or dispensers).

Automation level increases ranging from the "picker-to-parts" system to the "completely automated picking" one.

As stated above, this classification does not consider the single equipment type to be used. For example, in the "parts-to-picker" system it is possible to use both vertical storage carousel and miniload (*i.e.* an AS/RS designed for the storage and the retrieval of items, stored in modular storage drawers or bins). In the next sections each OPS will be briefly described on the basis of the resource requirements (labour, space and capital), the equipment types and main design issues. The "completely automated picking" system will not be considered, since it is employed in very limited contexts (*e.g.* drugs distribution). The focus will be mainly on systems with higher automation level: "pick to box", "pick and sort" and "parts to picker".

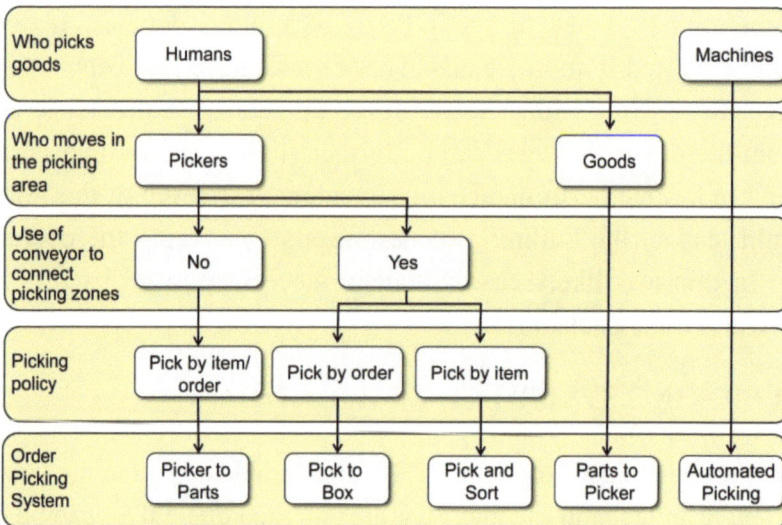

Figure 2: Classification of OPSs.

3.1. "Picker-to-Parts" System

"Picker-to-parts" system represents the very large majority of OPSs [1]. For example in Italy it has been implemented in over 50% of warehouses. It can be considered as the first choice for the picking activity. In such a system pickers travel (walking or driving a truck) from location to location, completing a single order or a batch of multiple orders, depending on the order picking policy. In the batch picking policy, the picked items are immediately sorted by the picker (sort-while-pick approach). Two types of "picker-to-parts" systems can be distinguished: low-level and high-level picking. In low-level order picking systems, items are picked from picking locations (*e.g.* racks, gravity flow racks, shelving) placed to the lower levels, with minimum vertical movement of pickers. The second, also called man-on-board OPS, employs high storage racks in which picking locations are visited by pickers on forklifts. In this OPS, the picking area is usually separated from the storage area. In such a way picking activity is conducted in a smaller area as compared to the storage one, with an increase of productivity. In the "picker-to-parts" system further optimisation can be carried out by means of certain levers, such as routing algorithms, items allocation policies and paperless operations that use radio frequency or voice picking devices. Therefore the picking productivity also strongly depends on the utilisation of the aforementioned optimisation levers.

3.2. "Pick-to-Box" System

In a "Pick-to-box" system (also known as "pick-and-pass" system) the picking area is divided into zones (also called picking stations). Each zone is assigned to different pickers. Items, once retrieved, are placed in bins that are transported *via* a conveyor line towards next picking station. Each bin corresponds (partially or completely) to a customer order ("order picking" policy). Each bin stops off in the zones where at least one SKU has to be picked. In each zone, after having picked all the required items, the picker pushes the bin back onto the main conveyor and the bin is thus moved to the following zone. The described process sequence is known as 'progressive zoning', whereby orders are picked sequentially zone-by-zone.

As an example in Fig. **3** the typical layout for the considered system is reported, assuming that items are grouped into two categories [5]. For each item class, a different storage solution (*i.e.* gravity flow racks, bin shelves) is used, according to the FRP (Forward Reserve Problem) approach. In particular, gravity flow racks are suitably used for high-volume (*i.e.* class 1) items, whereas bin shelves are used to store low-volume (*i.e.* class 2) items. The gravity flow racks are placed over the conveyor to increase picking productivity. The class 2 item area consists of a number AZ_j of picking aisles for each picking zone. In order to minimise the average travel distance, the depot position is located in the middle of each zone. The service time for a bin at each picking station consists of several components: setup time (*e.g.* time for checking the picking list, shifting the bins, labelling),

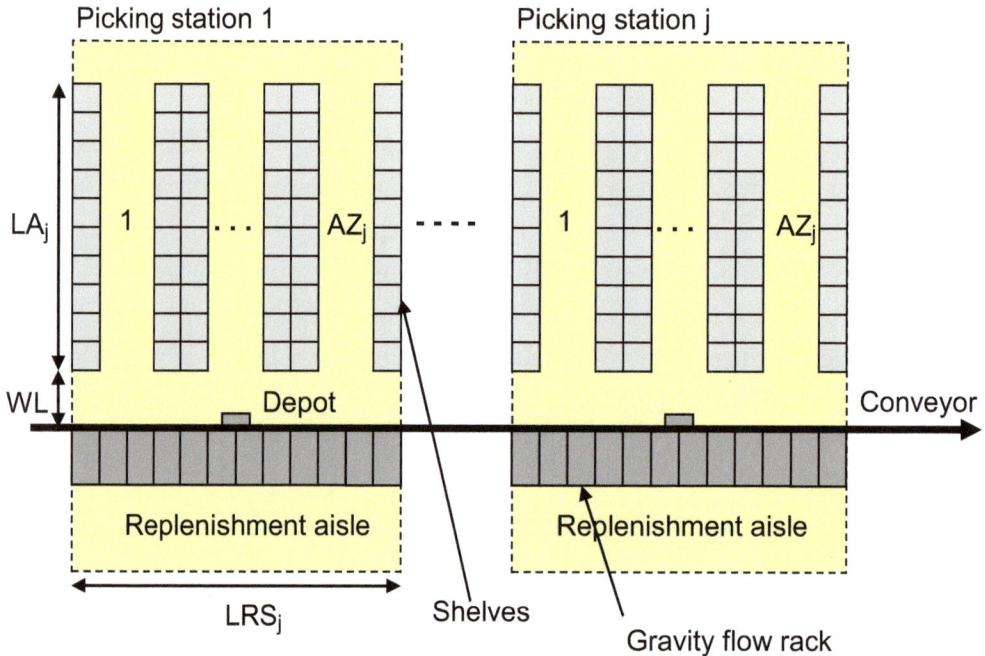

Figure 3: Example of a layout of the pick-and-pass system where AZ is the number of aisles in each picking station, LRS_j is the width of each picking station, LA_j is the length of each picking aisle, WL is the width of cross aisle.

travel time, and the picking time for order lines [6]. Travel time depends on the number of order lines to be picked at the picking station, the location of these order lines in the picking station, and the travel speed of pickers. The design of a pick-and-pass system is extremely specific due to the division of the picking area into zones [7]. A key design factor is the definition of the number and size of the zones in such a way that minimises the total order picking cost. For a fixed picking area size, larger picking zones (and hence a smaller number of zones) reduce picking productivity, thus increasing the service time per zone due to the longer travel time in zones. On the other hand, smaller and hence more numerous zones increase picking productivity, but decrease the use of pickers, while enhancing the effect of order setup time (*i.e.* the order tends to be split into more than one station) and the costs for each station (*e.g.* deviators, identification and accumulation systems). The performances of the OPS are mainly function of two main design parameters: order size and number of items [5]. As the number of items increases, the advantage of separating the picking area into a larger number of zones increases. As the order size increases the annualised costs fall, whereas the average throughput time tends to rise. As a matter of fact, if the same number of order lines are split into a smaller number of larger orders, the impact of the variable times on each order increases compared to the impact of the fixed times. As a consequence, the marginal benefit due to a reduction in travel (*i.e.* obtained by sub-dividing the picking area into a larger number of zones) is emphasised. This effect is reduced when the size of the picking zone becomes smaller, thus determining a higher impact of fixed times. As such, the pick-and-pass system may be potentially more suitable as the order size increases, compatibly with the bin capacity.

3.3. "Pick-and-Sort" System

In "pick-and-sort" system picks are grouped in picking waves, where for each single item the quantity resulting from the batching of multiple orders is retrieved. After retrieval, items are placed onto a takeaway conveyor connecting the picking area to the sorting system. As shown in Fig. **4**, at the end of the takeaway conveyor there is a closed-loop conveyor, with automatic divert mechanisms and accumulation lanes. A computerised system determines the destination sorting lane (also called chute) for each item. Each sorting lane refers to one or sometimes more individual customer orders. Upon reaching the appropriate sorting lane, packers fill the shipping boxes which are subsequently taken to the dispatch area.

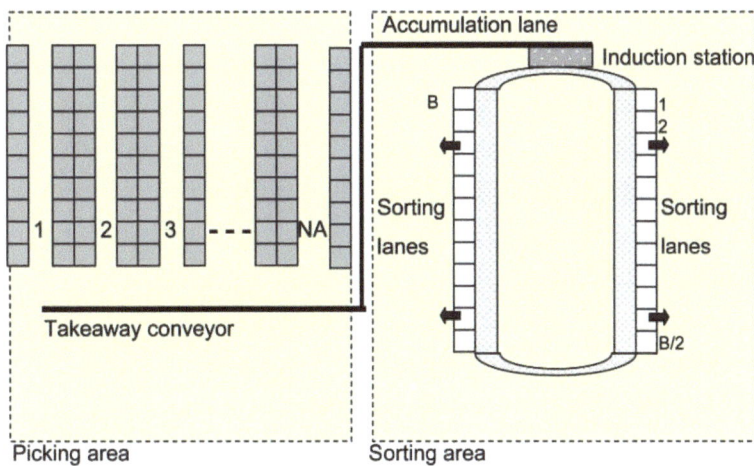

Figure 4: Pick-and-sort layout (top view), where NA is the number of aisles in picking area and B the number of sorting lanes.

There are, therefore, two main areas in a pick-and-sort system: the picking area and the sorting area. The sorting area, in turn, comprises 3 sub-components: the merge subsystem, the induction subsystem, the sorting subsystem. The picking area (also called the forward area) is no different from other picking systems. For instance, the picking area can be represented by pallet racks, with the racks at the lowest levels dedicated to picking activities. The merge subsystem (also called the accumulation subsystem or accumulation lane) is the conveyor that collects items from picking and receiving, and accumulates, merges and prepares them for sorting. The objective is to provide a constant flow of items to the sorting system. The conveyor carries out the accumulation function between the execution of one wave in the picking area and a subsequent wave in the sorting subsystem. Often the accumulation lane is fed by a takeaway conveyor, which, in turn, is close to the aisles of the picking area. As a consequence each picker can operate in just a small part of the picking area. Induction subsystems are designed to place items onto the sorter at the correct rate and with the required gap and package orientation to ensure the proper operation of the sorter. Induction can be performed both manually and automatically. Several, different technologies are available for automated sorting systems (*e.g.* bomb-bay, tilt-tray, cross-belt sorters). The choice between the different discharge mechanisms is based on the physical dimensions of the items to be sorted and is influenced by factors such as the weight, size, and fragility of the items. As a result of customer order aggregation, the same item may be picked, thus reducing the number of different locations to be reached (overlapping effect). Therefore, the organisation of the pick-and-sort system is based on the separation of the daily picking activities into picking waves, where the wave length is defined as the period of time in which a group of orders is processed in one area of the OPS (*i.e.* picking area) before proceeding to the next area (*i.e.* sorting system). The wave length has an effect on both picking efficiency and sorting costs. Longer waves typically lead to greater picking efficiency as a result of the shorter average distance between two picks (pick density) and the greater overlapping of items. However, longer wave durations also lead to increased costs due to more work-in-process, which then requires additional accumulation space and more sorting equipment. The productivity of the "pick-and-sort" system is higher than the one usually measured within "picker-to-parts" systems. Since the storage locations are visited less frequently, pickers travel time is reduced, above all if pickers operate in a small part of the forward area (zoning policy). When designing a "pick-and-sort" system great attention must be paid on trade-offs between picking and packing efficiencies. In addition, the trade-off between the capital costs involved with implementing an automated sorter and the manual labour saving must be carefully weighed. This system seems to be preferable in case of high overlapping of order lines, a high outflow and absence of brittle products [3].

3.4. "Parts-to-Picker" System

In "parts-to-picker" system the product totes are moved from the high-density storage area to the picking bays (also called work stations), where the pickers select the required amount of each item and they place it on a shipping unit load (also called order tote). Afterward, the product totes, if not empty, are transported back to the storage area. As described in Fig. **5**, the linkage between the storage area and picking bays can be carried out by an automatic device, such as a conveyor. In the storage area, different types of equipment may be implemented: carousels, modular vertical lift modules, miniloads, and automated storage and retrieval systems (AS/RS) using pallets as unit load. The advantage of this system derives from a better space utilization and above all the reduction of picking cost (*i.e.* in terms of labour required). Nevertheless, this system can be subject to wasted labour as picking operators can find themselves waiting for items to be delivered to their picking location, as a consequence of bottlenecks in feeding the picking bays. This OPS seems to be preferable in case of a large number of items and a relatively large outflow [3]. Recently, material handling suppliers have proposed a new solution for automating the handling of goods: Autonomous Vehicle based Storage and Retrieval System (AVS/RS). This solution has been exploited in recent warehouse implementations when the "parts to pickers" systems are concerned. AVS/RS can be seen as an evolution of AS/RSs. In AS/RS unit loads are handled using aisle-captive storage cranes that move simultaneously in the vertical and horizontal dimensions. Conversely, on the other hand in AVS/RS unit loads are handled by vehicles moving horizontally along rails within storage racks with vertical movement provided by lifts mounted along the rack periphery (Fig. **6**).

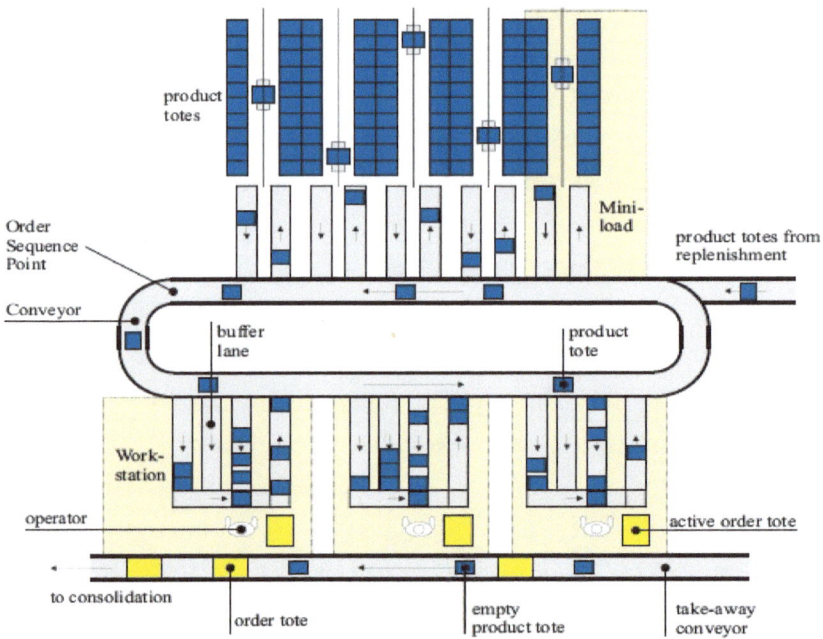

Figure 5: Parts to picker layout (top view) [8].

Depending on the unit load type, the number of vehicles per level and the ratio between the number of vehicles and the number of lifts are subject to change. When the unit load is constituted by a pallet, the solution most often implemented is represented by "tier to tier" AVS/RS, *i.e.* the vehicles shifts among the different level of the same rack. When the unit load is constituted by a product tote, the solution most often implemented is represented by "tier captive" AVS/RS, where each single vehicle is dedicated to fulfill the requirements of a single storage level [9]. This is the most performing solution (especially when one vehicle per aisle is employed), but the most expensive one due to the large number of employed vehicles. For "parts to picker" OPS, the "tier captive" AVS/RS solution potentially turns to be the most interesting one for its high throughput capacity. In this way it is potentially possible to reduce bottlenecks in feeding the picking bays. Finally the expected AVS/RS energy consumption is lower than in AS/RS, as horizontal movements are performed by vehicles, which are lighter than cranes. Therefore the AVS/RS can improve Supply Chain Sustainability [10].

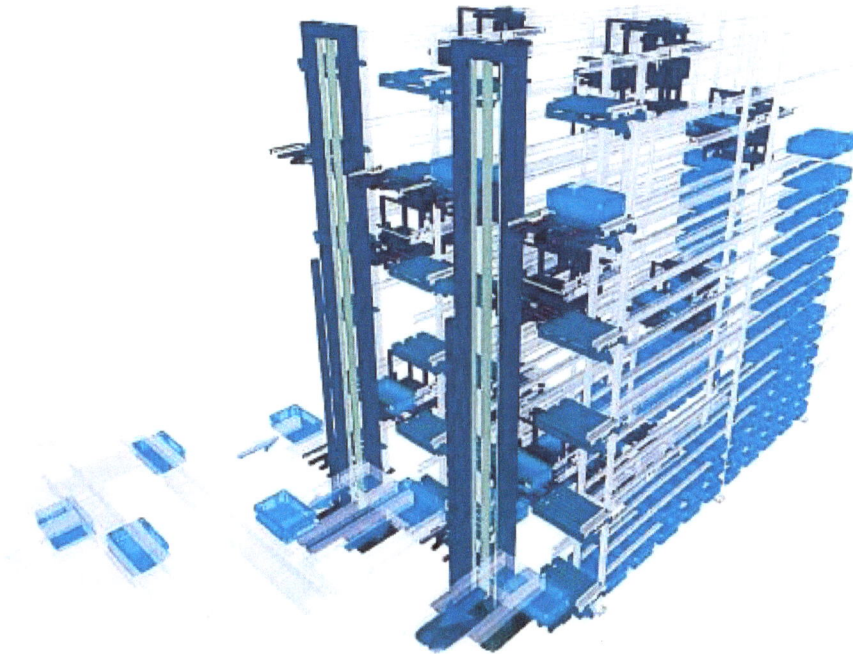

Figure 6: Illustration of AVS/RS (Quickstore by Vanderlande picture).

4. MAIN CRITERIA TO IDENTIFY THE RIGHT OPS

As shown in the previous sections the choice of right OPS is a challenging task. Analyzing a significant data set of OPSs, a recent study [3] has shown, that the OPS identification is mainly function of 3 drivers:

- order size (expressed in terms of cubic volume),

- number of SKUs and

- daily activity (expressed in terms of average number of daily order lines).

Figs. **7** and **8** depict the picking systems analyzed by [3]. The analysed OPS have been plotted on a scatter diagram, built considering the number of items on the x axis and the daily activity (expressed as number of order lines per day) on the y axis. Because of the high data concentration, the logarithmic scale has been used for both dimensions. Fig. **7** considers OPSs that fulfil "small orders" (*i.e.* the average order size, expressed in volume, is lower than (or equal to) 0.5 m^3). Fig. **8** considers OPSs that fulfil "large orders" (*i.e.* the average order size, expressed in volume, is higher than 0.5 m^3).

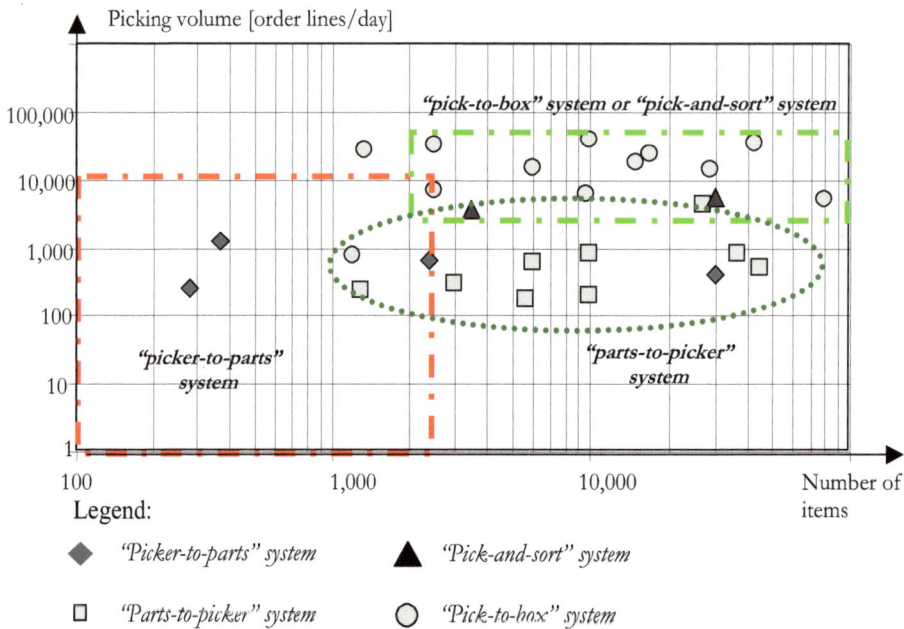

Figure 7: OPS matrix, where small orders are concerned [3].

In case of small orders (as shown in Fig. **7**), where the number of retrieval operations (order lines picked per day) and number of items are large, it seems that the most suitable solution could be "pick-to-box" system or "pick-and-sort" system. The difference between these OPS groups is partially related to the overlapping effect, in turn depending on the number of managed items. Indeed, as the number of items increases, the "pick-and-sort" system is less performing than "pick-to-box" because of the reduced overlap among order lines. Furthermore the "pick-to-box" system is more suitable for the retrieval of small items [3]. Conversely, reducing the number of order lines picked per day, adoption of a "picker-to-parts" system appears suitable, especially when the number of SKUs is low (approximately less than 1,000). The "parts-to-picker" system represents an intermediate alternative, since its application field is identified by a significant number of items (approximately higher than 1,000) and a relatively small number of order lines (approximately about 1,000-2,000 order lines per day). As shown in Fig. **8**, in case of large orders "picker-to-parts" seems to be the most suitable system. According to [3] the higher the order size, the more significant the consequent advantage in adopting this system. In fact a large order size allows to apply optimisation approaches, such as the zone-picking policy. However with large orders as well the "parts-to-picker" system could be suitable in case of a large number of items (approximately higher than 1,000) and relatively small of order lines (approximately 1,000-2,000 order lines per day).

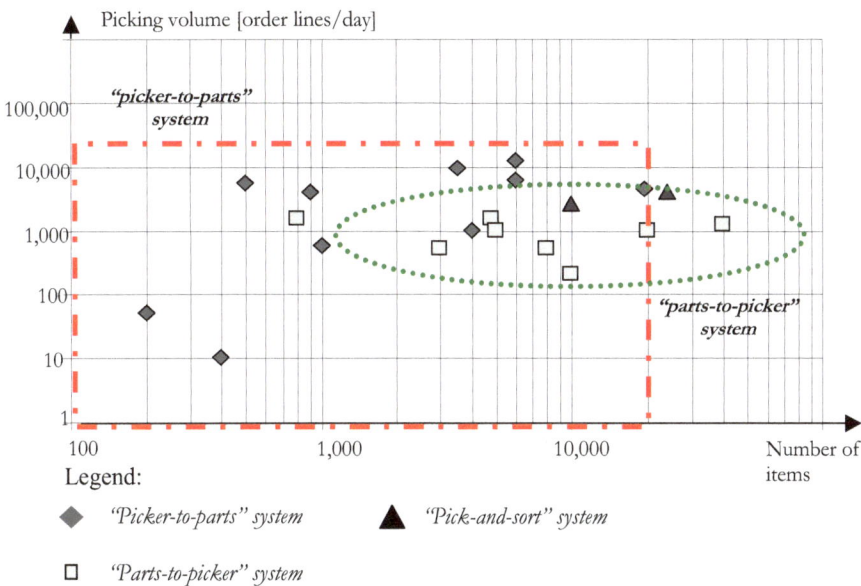

Figure 8: OPS matrix, where large orders are concerned [3].

Besides the aforementioned aspects, (*i.e.* order size, number of SKUs, daily activity), in the selection of OPSs other factors must be taken into account, which are harder to generalise, such as: risk attitude of the company (*i.e.* payback time), variability of business processes, unit labour cost, flexibility of manpower. For instance the "picker-to-parts" system is the most widespread system across 3PLs (Third Party Logistics), because it allows a remarkable operating flexibility, well-fitting for the short average outsourcing contract length. Finally, while choosing OPS systems, a relevant factor to be considered is represented, as abovementioned, by the overlapping effect. As previously stated, the greater the overlapping effect, the greater the potential suitability of "pick and sort" systems. The a priori quantification of the overlapping effect, *e.g.* during the selection stage, turns to be significantly difficult. Assuming random batching of orders (a frequent condition when, for example, delivery scheduling needs to be respected and orders can therefore not be batched according to any other criteria) and demand for any order line independent, according to [11] it is possible to obtain an analytic estimate of the overlapping effect for a generic batch of orders (called picking wave). Taking LW_{max} as the maximum number of order lines to pick per picking wave (equal to the sum of the order lines on the orders processed in the wave), the expected number of items to pick per wave (LW) has the form:

$$LW = \sum_{j=1}^{S}\left[1-(1-\frac{1}{S})^{LW_{max}}\right] = S \cdot \left[1-(1-\frac{1}{S})^{LW_{max}}\right] \tag{1}$$

where:

S indicates the overall number of items in the picking area.

$\dfrac{1}{S}$ indicates the probability that item j is picked, assumed to be equal for all items.

$(1-\dfrac{1}{S})$ indicates the probability that item j is not picked.

$(1-\dfrac{1}{S})^{LW_{max}}$ indicates the probability that item j is not picked in the picking wave with LW_{max} order lines.

$1-(1-\dfrac{1}{S})^{LW_{max}}$ indicates the probability that item j is picked (once or more) in the picking wave with LW_{max} order lines.

Considering a general case in which the items (S) are divided into i classes (with S_i items per class), the total number of items to be picked (LW) is computed by summing the contribution of each class i (LW_i). Therefore, LW may be estimated as follows:

$$LW = \sum_{i=1}^{I} LW_i \tag{2}$$

Similar to (1), LW_i is given by:

$$LW_i = \sum_{j=1}^{S_i} \left[1 - (1 - \frac{1}{S_i})^{LW_{max,i}} \right] = S_i \cdot \left[1 - (1 - \frac{1}{S_i})^{LW_{max,i}} \right] \tag{3}$$

Where $LW_{max,i}$ is the maximum number of order lines of the ith class picked per wave. This value may be estimated as the product of LW_{max} and the order frequency of product class i (f_i).

According to [11] the expressions (1-3) to estimate LW present a low margin of error. From expressions (2) and (3), it is possible to assess the overlapping effect as a function of the operating conditions. As an example, Table **1** reports the expected number of items to be picked per picking wave in absolute (LW) and percentage terms compared to the total number of lines requested per wave, considering 25 different operating conditions. The operating conditions depend on the ABC curve and on the number of Stock Keeping Units. The ABC curve assumes values equal to 50/20, 60/20, 70/20, 80/20 and 90/20, while the number of SKUs varies in a range comprised between 4,000 and 20,000 with intervals of 4,000. For instance, in the case of a 50/20 curve, class 1 items will have S_i equal to 20% of the items (S) and LW_{max} equal to 50% of LW_{max}. Class 2 will be composed of the remaining items, from which the remaining number of order lines will be picked. The other assumptions included in Table **1** are: order size (10 order lines per order), wave size (250 orders), order lines per wave, *i.e.* LW_{max} (2,500). Results have been extended increasing the order size from 10 to 20 lines (Table **2**). The results reported in Tables **1** and **2** show:

- As the number of items increases, the overlapping effect declines.

- As the ABC curve becomes more skewed, the overlapping effect grows (with a constant number of items, the difference between the 50/20 and the 90/20 ABC curve is at least 20%), as the picks are concentrated on just a few items. This effect is greater if there is a small number of items.

As the overall number of lines to pick in a wave increases, the overlapping effect rises in percentage terms (*e.g.* for a 50/20 ABC curve with 4,000 SKU, it changes from picking 66.7% of the 2,500 potential lines requested to picking just 50% of the 5,000 lines requested).

Table 1: Expected number of items to be picked per picking wave in absolute and percentage terms compared to the total number of lines requested per wave (2,500), as a function of the number of items and the ABC curve.

Number of SKU	ABC curve				
	50/20	60/20	70/20	80/20	90/20
4,000	1,667 (66.7%)	1,536 (61.5%)	1,379 (55.2%)	1,197 (47.9%)	993 (39.7%)
8,000	2,003 (80.1%)	1,890 (76%)	1,772 (70.9%)	1,623 (64.9%)	1,453 (58.1%)
12,000	2,147 (85.9%)	2,065 (82.6%)	1,964 (78.6%)	1,844 (73.8%)	1,707 (68.3%)
16,000	2,226 (89%)	2,160 (86.4%)	2,077 (83.1%)	1,978 (79.1%)	1,864 (74.5%)
20,000	2,276 (91%)	2,220 (88.8%)	2,150 (86%)	2,066 (82.7%)	1,969 (78.8%)

Table 2: Expected number of items to be picked per picking wave in absolute and percentage terms compared to the total number of lines requested per wave (5,000), as a function of the number of items and the ABC curve.

Number of SKU	ABC curve				
	50/20	60/20	70/20	80/20	90/20
4,000	2,500 (50%)	2,268 (45.4%)	1,988 (39.8%)	1,653 (33.1%)	1,260 (25.2%)
8,000	3,334 (66.7%)	3,073 (61.5%)	2,758 (55.2%)	2,395 (47.9%)	1,985 (39.7%)
12,000	3,754 (75.1%)	3,518 (70.4%)	3,231 (64.6%)	2,897 (57.9%)	2,519 (50.4%)

Table 2: contd....

16,000	4,006 (80.1%)	3,799 76%)	3,544 (70.9%)	3,245 (64.9%)	2906 (58.1%)
20,000	4,174 (83.5%)	3,991 (79.8%)	3,765 (75.3%)	3,498 (70.0%)	3,194 (63.9%)

ACKNOWLEDGEMENT

Declared none.

CONFLICT OF INTEREST

The author(s) confirm that this chapter content has no conflict of interest.

DISCLOSURE

Part of information included in this chapter has been previously published in: Dallari, F., Marchet, G., Melacini, M. (2009), "Design of order picking system". International Journal of Advanced Manufacturing Technology Vol. 42, No. 1-2, pp. 1-12.

REFERENCES

[1]   De Koster, R., Le-Duc, T., Roodbergen, K. (2007), "Design and control of warehouse order picking: a literature review", *European Journal of Operational Research*, Vol. 182, No. 2, pp. 481-501.

[2]   Baker, P. and Halim, Z. (2007), "An exploration of warehouse automation implementations: cost, service and flexibility issues", *Supply Chain Management: an International Journal*, Vol. 12, No. 2, pp. 129-138.

[3]   Dallari, F., Marchet, G., Melacini, M. (2009), "Design of order picking system", *International Journal of Advanced Manufacturing Technology*, Vol. 42, No. 1-2, pp. 1-12.

[4]   Van der Berg, J.P. (1999), "A literature survey on planning and control of warehousing systems", *IIE Transactions*, Vol. 31, No. 8, pp. 751-762.

[5]   Melacini M., Perotti, S., Tumino, A. (2011), "Development of a framework to design Pick-and-pass Order picking System", *International Journal of Advanced Manufacturing Technology*, Vol. 53, No. 9, pp. 841-854.

[6]   Yu M, De Koster R (2008), "Performance approximation and design of pick-and-pass order picking systems", IIE Transactions, Vol. 10, No. 11, pp. 1054-1069

[7]   Petersen, C.G. (2002), "Considerations in order picking zone configuration", *International Journal of Operations and Production Management*, Vol. 22, No. 7, pp.793-805.

[8]   Andriansyah A. R., de Koning W.W.H., Jordan R., Etman L.F.P., Rooda J.E. (2008), *Simulation Study of Miniload-Workstation Order Picking System*, SE Report: Nr. 2008-07, ISSN: 1872-1567

[9]   Tappia, E., Marchet, G., Melacini, M., Perotti, S. (2012), "Analytical model to estimate performances of autonomous vehicle storage and retrieval systems for product totes", *International Journal of Production Research*, Vol. 50, No. 24, pp. 7134-7148.

[10]   Colicchia, C., Melacini, M., Perotti, S. (2011), "Benchmarking supply chain sustainability: Insights from a field study", *Benchmarking: an international journal*, Vol. 18, No. 5, pp.705-732.

[11]   Marchet, G., Melacini, M., Perotti S. (2011), "A model for design and performance estimation of pick-and-sort order picking systems", *Journal of Manufacturing Technology Management,* Vol 22, No 2, pp. 261-282.

Send Orders for Reprints to reprints@benthamscience.net

CHAPTER 3

Comparison of Communication with the CNC Systems

Yusri Yusof* and Noordiana Kassim

Faculty of Mechanical and Manufacturing Engineering, Universiti Tun Hussein Onn, Malaysia

Abstract: This book chapter gave out explanation of ISO 14649 or STEP-NC and ISO 6983 or G&M Code. The structures of both standards were being explained and comparison between them were being made and studied into details through research done on both codes representation. Summary of the comparison were being groups into specific categories that gave out details view on how ISO 14649 are better equipped as compared to ISO 6983. The comparison made was on the general main categorization of the differences rather than focusing on the difference in the details structure of the actual codes and was based on the intention of highlighting the benefits of ISO 14649 instead of degrading ISO 6983. In general, both codes have its own advantages but as we move towards the future, requirements and advancement for manufacturing sectors evolved with time and its vital for industries to adapt to these changes to provide better end products, and *via* this one to one comparison, the strength of STEP-NC or ISO 14649 can be further highlighted.

Keywords: STEP-NC, ISO14646, ISO 6983, Computerised Numerical Control (CNC), Computer Aided Design (CAD), Computer Aided Manufacture (CAM), G and M code, data flow, product data, information loss, interoperability & traceability.

INTRODUCTION

The objective of this section is to study Numerical Control fundamentals issues and its relationship with G & M code and STEP-NC in order to determine the factors that differentiate the two Numerical Control codes that were introduced by ISO. An overview of NC system will be explained further in this chapter that will help understands the basic concept of NC systems. A background research on G & M code and STEP-NC will be reviewed and lastly a one to one comparison between both codes will be looked into in a more in depth scenario.

*__Address correspondence to Yusri Yusof:__ Faculty of Mechanical and Manufacturing Engineering, Universiti Tun Hussein Onn, Malaysia; Tel: 07-453 7700; Fax: 07-453 6080; E-mail: yusri@uthm.edu.my

HISTORY OF NUMERICAL CONTROL

According to Gibbs [1, 2], many factors contribute to this economic viability. The dramatic effect of numerical control on traditional engineering production techniques is now well appreciated. Machines controlled in this way are capable or working for many hours every day virtually unsupervised. They are readily adaptable to facilitate production of a wide range of components. Every function traditionally performed by the operator of a standard machine tool can be achieved *via* a numerical control machining program [2].

The early NC machines and today's CNC utilize the same standard of programming, namely G & M Code formalized as the ISO 6893 standard. Most Computer Numerical Control (CNC) machines are programmed in the ISO 6983 "G-code" language [3], a standard developed and effectively implemented by many CNC machines back in the 1950s through to 1970s [4]. The background of the CNC equipment at that time is much like that of the computer systems in that era where punched cards and tapes were the main information medium and the computing power was only a tiny fraction of what the current computers are offering [4].

According to Newman [5], starting in the 1970s, significant development has been made towards automatic and reliable CNC machines with new processes such as punching and nibbling, laser cutting, and water jet cutting, which are now common place. The invention of minicomputers, and later microcomputers, has brought a massive improvement in the capabilities of CNC machines with the ability of multi-axis, multi-tool, and multi-process manufacturing.

ISO 6983 (G & M CODE)

ISO 6983 or better known as G & M code is a list of instructions that has been used for decades to communicate with the CNC systems. In actuality, G-codes are only a part of the NC programming language that controls NC and CNC machine tools. This standard goes back to the time of punched cards and does not match to the level of advancement that the current technologies have to offer. In G-codes, codes are divided into groups or families to distinguish which codes can function simultaneously in a program.

Most CNC machines are programmed in the ISO 6983 "G and M code" language [6, 7]. Fig. **1** demonstrates the G & M code in actual G & M code programming environment, which is the G & M code part program for pocket process.

```
G0 G90 X88. Y33.                                                    G0 Z-4.
G43 H0 Z25. T1                                                   X88. Y33.
Z5.                                                                 Z-24.
G1 Z-1. F3.                                                      G1 Z-30.
X51.                                                                 X51.
X50.243 Y34.757                                          X50.243 Y34.757
X51. Y33.                                                       X51. Y33.
Y27.                                                                 Y27.
X50.243 Y25.243                                          X50.243 Y25.243
X51. Y27.                                                       X51. Y27.
X88.                                                                 X88.
X89.757 Y25.243                                          X89.757 Y25.243
X88. Y27.                                                       X88. Y27.
Y33.                                                                 Y33.
X89.757 Y34.757                                          X89.757 Y34.757
X91.5 Y37.5                                                  X91.5 Y37.5
X47.5                                                               X47.5
X45.743 Y39.257                                          X45.743 Y39.257
X47.5 Y37.5                                                  X47.5 Y37.5
Y21.5                                                               Y21.5
X45.743 Y20.743                                          X45.743 Y20.743
X47.5 Y21.5                                                  X47.5 Y21.5
X91.5                                                               X91.5
X94.257 Y20.743                                          X94.257 Y20.743
X91.5 Y21.5                                                  X91.5 Y21.5
Y37.5                                                               Y37.5
X94.257 Y39.257                                          X94.257 Y39.257
X100. Y46.5                                                  X100. Y46.5
X40.                                                                 X40.
G3 X38.5 Y45. R1.5                                      G3 X38.5 Y45. R1.5
G1 Y15.                                                          G1 Y15.
G3 X40. Y13.5 R1.5                                      G3 X40. Y13.5 R1.5
G1 X100.                                                        G1 X100.
G3 X101.5 Y15. R1.5                                    G3 X101.5 Y15. R1.5
G1 Y45.                                                          G1 Y45.
G3 X100. Y46.5 R1.5                                    G3 X100. Y46.5 R1.5
G0 Z24.                                                          G0 Z-5.
X81. Y25.                                                       X81. Y25.
Z5.                                                                 Z-24.
G1 Z-1.                                                          G1 Z-30.
Y37.5                                                               Y37.
G3 X70. Y49. R11.                                        G3 X70. Y49. R11.
G1 X40.                                                          G1 X40.
G3 X36. Y45. R4.                                          G3 X36. Y45. R4.
G1. Y15.                                                        G1. Y15.
G3. X40. Y11. R4.                                        G3. X40. Y11. R4.
G1. X100.                                                      G1. X100.
G3 X104. Y15. R4.                                        G3 X104. Y15. R4.
G1. Y45.                                                        G1. Y45.
G3. X100. Y49. R4.                                      G3. X100. Y49. R4.
G1. X70.                                                        G1. X70.
G3. X58. Y37. R11.                                      G3. X58. Y37. R11.
G1. Y25.                                                        G1. Y25.
G0. Z24.                                                        G0 Z25.
X88. Y33.                                                   G91 G28 Z0 M5
...                                                                  G64
```

Figure 1: ISO 6983 programming for pocket process [8].

ISO 10303 (STANDARD FOR THE EXCHANGE OF PRODUCT MODEL DATA)

For mechanical parts, the description of product data has been standardized by ISO 10303. This leads to the possibility of using standard data throughout the entire process chain in the manufacturing enterprise. Impediments to realize this principle are the data formats used at the machine level [6].

ISO 10303 is an ISO standard for the computer-interpretable representation and exchange of product manufacturing information. Its official title is "*Industrial automation systems and integration - Product data representation and exchange*", known as "*STEP*" or "*Standard for the Exchange of Product model data*" [10]. The International standard's objective is to provide a mechanism that is capable of describing product data throughout the life cycle of a product, independent from any particular system. The nature of this description makes it suitable not only for neutral file exchange, but also as a basis for implementing and sharing product databases and archiving.

In design and manufacturing, many systems are used to manage technical product data. Each system has its own data formats; as a result the same information has to be entered multiple times into multiple systems leading to redundancy and errors. The problem is not unique in manufacturing but more acute because design data is complex and 3D leading to increase scope for errors and misunderstandings between operators. The National Institute of Standards has estimated that data incompatibility is a 90 billion dollar problem for manufacturing industry [11]. The ultimate goal is for STEP to cover the entire life cycle, from conceptual design to final disposal, for all kinds of products. However, it will be a number of years before this goal is reached. The most tangible advantage of STEP to users today is the ability to exchange design data as solid models and assemblies of solid models [9].

Typically STEP can be used to exchange data between CAD, Computer-aided manufacturing, Computer-aided engineering and Product Data Management. STEP is addressing product data from mechanical and electrical design, Geometric dimensioning and tolerance, analysis and manufacturing, with additional information specific to various industries such as automotive, aerospace, building construction, ship, oil and gas, process plants and others.

Express is a standard data modeling language for product data. The description of process data is done using the EXPRESS language as defined in ISO 10303-11 [6]. EXPRESS-G is a standard graphical notation for information models. It is a useful companion to the EXPRESS language for displaying entity and type definitions, relationships and cardinality. EXPRESS-G diagrams are an aid for understanding large information models. The diagrams show relationships and structure more clearly than the plain EXPRESS text [9].

ISO 14649

Background

STEP-NC (ISO 14649) is a STEP-compliant data interface [11, 12]. The enhancements in open control coupled with modern computer technology have provided the basis to re-examine the way in which CNC machines have been programmed during the last 50 years. These enhancements, combined with the application of new STEP standards in manufacturing, have provided a valuable stimulus to raise the awareness of controller suppliers, CNC machine-tool manufacturers, and CAx software vendors to develop new CAM and CNC software products that have the possibility to bring about a new era for CNC manufacture in the 21st century [13].

Problems with G & M Code

The manufacturing environment changed, with more collaboration and intelligence since the 1990s. High-speed machining, high-precision machining and multi-axis complex machining have extensively enhanced the productivity and quality of manufacturing. Furthermore, advanced internet technology has introduced a new paradigm of e-Manufacturing. Because ISO 6983 was developed at a time when computer power was limited and machines were controlled offline, the needs and possibilities, then, were very different from those of today. These machines used simple instructions to move tools through the air and for cutting metal. According to Suh [14], the summary problems of G & M codes have been reported as follows:

- Information Loss

  A G & M-code part program is defined by simple alphabetical or numerical codes and delivers only limited information to the CNC (excluding valuable information such as part geometry and the process plan used to generate the NC code), it makes the CNC simply an executing mechanism, completely unaware of the motions being executed.

- Difficult Traceability

  As a G & M-code part program is made up of a coded set of numbers for axis movements, it is not easy for machine operators to understand the operational flow, machining condition and specification of tools only by reading the low-level part program. In particular, it makes it more difficult, not only in finding which part happens to cause problems, but also modifying the program for solving these problems.

- Lack of Interoperability

  The G & M code schema is dependent on the machine tool builder or controller maker. For this reason, it is necessary to apply a post-processing process in which the program is adapted to a specific CNC machine configuration. This post-processing is one of the main interferences with the seamless data flow in the CAD-CAM-CNC chain.

- Non-compatibility with higher level systems

  In higher-level manufacturing systems, such as office level CAD/CAM, compatible information exchange is being increased gradually by the introduction of STEP. In contrast, the rich information environment has almost perished at the CNC on the shop floor level. Also, there is little information feedback from the CNC, which makes the shop floor status obscure to the upper systems.

STEP-NC File Structure

To better explain the STEP-NC structure, based on diagram in Fig. **2**, a STEP-NC file consists of Header section and a Data section. The HEADER will contain general information like comments about the program, filename, author, date or organization. The DATA section are divided into three parts, (a) workplan and executables, where the STEP-NC will perform execution of the sequenced tasks listed; (b) technology and description, where STEP-NC provides description of the technological data like workingsteps, tools and technologies involved; (c) geometry description. Therefore, the CNC machines are provided with rich information for part manufacture and the user will access the system using graphic interface (shop floor programming).

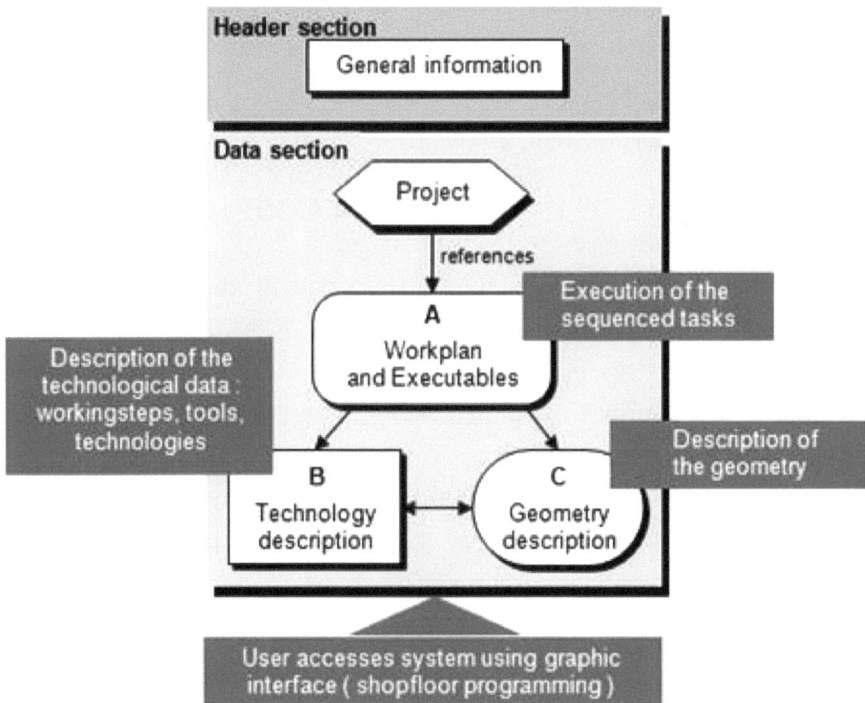

Figure 2: Part of STEP-NC structure format [15].

Fig. **3** is the graphical display of the workpiece that uses the STEP-NC code for pocket process.

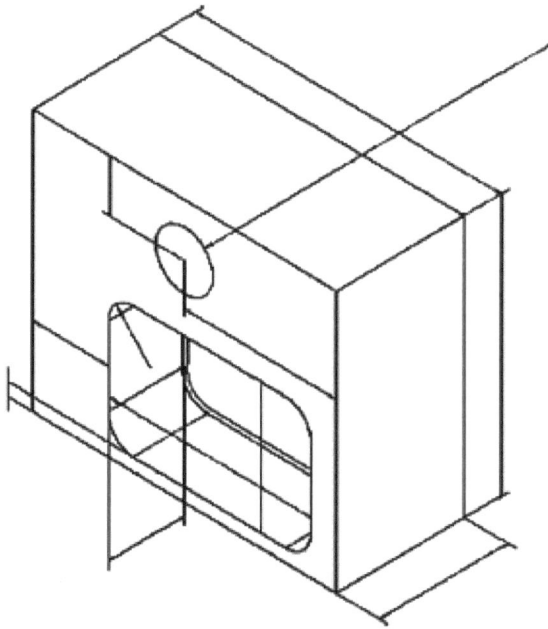

Figure 3: Example part for simple milling process [16].

COMPARISON BETWEEN ISO 6983 (G & M CODE) AND ISO 14649

Many papers have been published on the importance and the vast benefits of STEP-NC. Some papers did mention the fact that STEP-NC is better as compared to the 50 years old G & M Code standards and it is much better if the concept of G & M Code is being further explained and explored so readers and researchers could understand better the importance of STEP-NC. CNC technologies and development has been a vital part in the manufacturing process. G & M Code has been assimilated and associated with NC technologies for more than five decades and further and in-depth understanding of G & M Code itself would definitely help in spreading the idea and concept of STEP-NC.

Without proper study and understanding of G & M Code, replacing it with the brand new STEP-NC concept would be hard to accomplish. The individual mentioned flaw of G & M Code and in what way did STEP-NC overcome that flaw should be further studied and explained and be put on visual one-to-one comparison. This research would provide a visual one to one comparison between

G & M Code (ISO 6983) and STEP-NC Code (ISO 14649). The individual mentioned flaw of G & M Code and in what way STEP-NC overcame that flaw was put on a visual explanation and comparison.

Low Level Code *vs.* High Level Code

The term 'high level' does not imply that the programming code or language is superior to 'low level' programming, rather, 'high-level' here refers to the higher level of abstraction from machine language. In comparison to low level codes, high level code programming uses natural language element which will be easier to use, or be more portable across platforms. Such codes hide the details of computer operations. The hiding of details is generally intended to make the code user-friendly, as it includes concepts from the problem domain instead of those used by the machine.

Low level code programming is a language that provides little or no abstraction from a computer's instruction set architecture. It is more computer or hardware friendly rather than user-friendly and it requires memorizing or looking up numerical codes for every instruction that it used. Essentially, the lower you are, the closer you get to the machine. At the opposite end, high-level code offers a great deal of abstraction, and are then said to be 'far' from the machine [17].

According to W. Maeder [12], ISO 6983 uses low-level codes to describe tool movements (such as G01) and switching instructions (such as M5).

Generally, a specific code like G or M code that's tells the machine tool what type of action to perform, like movement or tools changed. These codes can only be understood by the machine as it is much similar to machine language and very difficult to read by non experts of the codes itself, as seen in example of G & M code part program as shown in Fig. **4**.

These type or kind of codes are very difficult to understand and modified for someone with limited knowledge of G & M Code and rather unreadable to the unknowledgeable human eye. In other words, it is not user friendly and more machines friendly.

G0 G90 X88 Y33

G0 means *Rapid positioning*

G90 means *Absolute programming for type B and C systems*

X88 means *X axis 88*

Y33 means *Y axis 33*

The whole line will carry the meaning of:

"Rapid positioning of machine to absolute positioning in X axis 88 and Y axis 33."

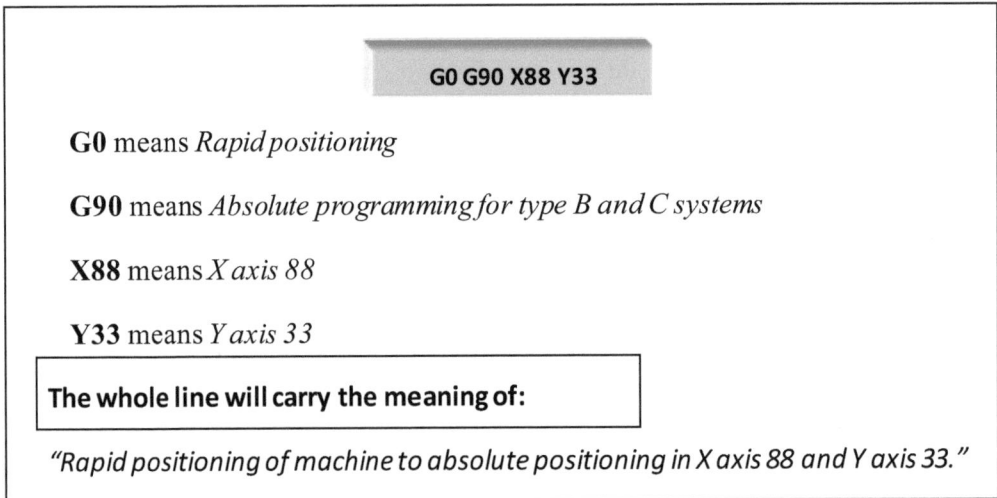

Figure 4: G & M Code Part Program Line.

STEP-NC codes are more user friendly in contrast to G & M Code. Each line represents the process that the code or program intends to perform rather clearly. As shown in Fig. **5**, instead of giving instructions in codes, STEP-NC gives instructions in meaningful words that can be easier understood and interpreted by others.

```
#1=PROJECT('EXECUTE EXAMPLE1', #2, (#4), $, $, $);

#2=WORKPLAN('MAIN WORKPLAN',(#10, #11, #12, #13), $, #8, $);

#4=WORKPIECE('SIMPLE WORKPIECE',#6, 0.010,$, $, $, (#68, #69));

#6=MATERIAL('ST-50','STEEL',(#7));
```

Figure 5: STEP-NC Codes Part Program.

The entity word **PROJECT** itself is self-explanatory to what it represents. Though, the sub-entities that contains in project need to be referred to, but it is readable to the unknowledgeable human eye, as it contains pointers of where the program should point or read next. The same situation applies to the entities words **WOKPLAN**, **WORKPIECE** and **MATERIAL** in the given example. Users can easily understand from the code that these codes will be using steel as its material. Fig. **6**, shows the architecture of low level codes and high level codes.

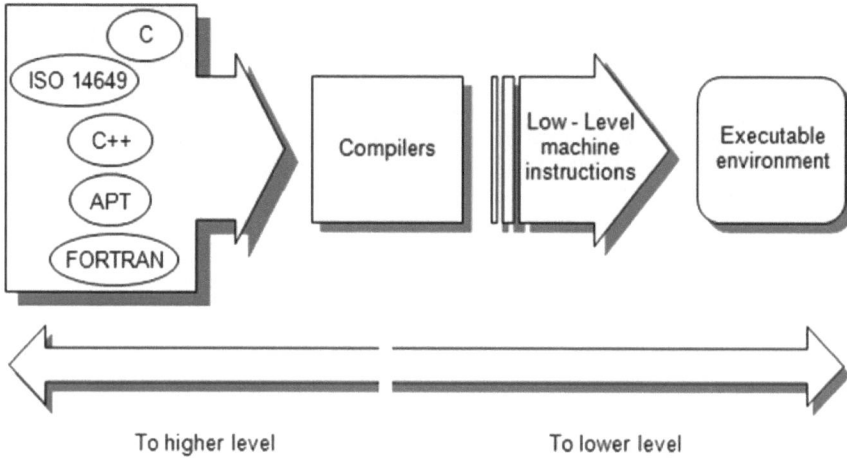

Figure 6: Low Level and High Level Components of Architecture [18].

Low Level Information *vs.* High Level Information

ISO 6983 or G & M Code is a low level language mainly specifying the cutter motion in terms of position and feed rate. It delivers only limited information to CNC and excludes the valuable information, such as part geometry and process plan implicated in the NC code [19]. The G & M code information available to a CNC machine is too low level in information, with which only minimum amount of optimization work can be carried out in real time or near real time.

```
G0 G90 X88 Y33

Z5

G1 Z-1 F3

X52

X50243 Y34757

X52 Y33

Y27

X50243 Y25243

X52 Y27
```

Figure 7: G & M Code Part Program.

A G & M Code are defined by simple alphabetical or numerical codes such as G, T, M, F, S, indicating the movement of a machine and an axis to the controller, which makes the CNC simply an executing mechanism, completely unaware of the motions being executed. For example, for these codes sequence in G & M Code in Fig. **7** not much information is being given or can be read from these codes. It only contains information to the machine that the program wants it to perform. What type of action to perform, to which what the alphabet G means or the axis movement of X, Y and Z. These show the low levelness of information in G & M Code.

With STEP-NC, both design and process planning information is available to a CNC machine. It is possible for the CNC machines, or their controllers, to perform high-level, intelligent activities, such as automatic part setup; automatic and optimal tool path generation; accurate machining status and result feedback; complete collision avoidance check; optimal workingstep sequence; adaptive control and on-machine inspection [7]. Fig. **8**, shows example of how high level information in STEP-NC are.

```
#6=MATERIAL('ST-50','STEEL',(#7));

#7=PROPERTY_PARAMETER('E=200000N/M2');

#8=SETUP('SETUP1',#71,#62,(#9));

#9=WORKPIECE_SETUP);

#10=MACHINING_WORKINGSTEP('WSFINISHPLANARFACE1',#62,#16,#19,$);

#11=MACHINING_WORKINGSTEP('WSDRILLHOLE1',#62,#17,#20,$);

#12=MACHINING_WORKINGSTEP('WSREAMHOLE1',#62,#17,#21,$);

#13=MACHINING_WORKINGSTEP('WSROUGHPOCKET1',#62,#18,#22,$);

#14=MACHINING_WORKINGSTEP('WSFINISHPOCKET1',#62,#18,#23,$);

#16=PLANAR_FACE('PLANAR FACE1',#4,(#19),#77,#63,#24,#25,$,());

#17=ROUND_HOLE('HOLE1 D=22MM',#4,(#20,#21),#81,#64,#58,$,#26);

#18=CLOSED_POCKET('POCKET1',#4,(#22,#23),#84,#65,(),$,#27,#35,#37,#28);

#19=PLANE_FINISH_MILLING($,$,'FINISHPLANARFACE1',10.000,$,#39,#40,#41,$,#60,#61,#42,1.500,$);

#20=DRILLING($,$,'DRILL HOLE1',10.000,$,#44,#45,#41,$,$,$,$,$,#46);

#21=REAMING($,$,'REAMHOLE1',10.000,$,#47,#48,#41,$,$,$,$,$,#49,.T.,$,$);
```

Figure 8: STEP-NC Part Program.

From sample of STEP-NC codes in Fig. **9**, one can draw out the information about the material from the MATERIAL entity, property parameter from the PROPERTY_PARAMETER entity , the machining workingstep involved from the MACHINING_WORKINGSTEP entity, the kind of features involved from the ROUND_HOLE entity (which explained that the codes is for developing a round hole inside a workpiece), as well as the process involved in making the features, like the drilling, reaming, planar face and plane finish millings from the entity DRILLING, REAMING, PLANAR_FACE and PLANE_FINISH_MILLINGS. The information provided by the STEP-NC codes is high level enough to be understood by the user.

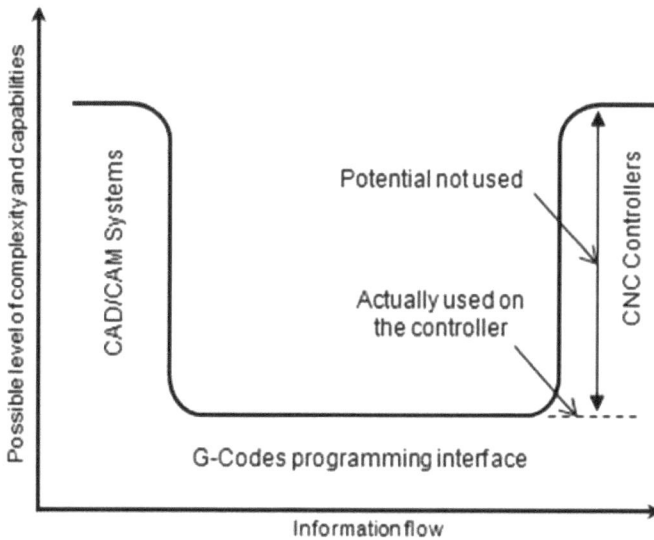

Figure 9: Impact of Using the G Code Programming Interface [18].

Simple Geometries *vs.* Complex Geometries

G-code focuses on the path of the cutter location (CL) and lacks most product and process information, such as the geometrical and topological information, part materials, tolerance, *etc.* [20]. The limited information in G-codes, makes it not sufficient to support complex machine functionality as it cannot perform feature, process and tool recognition. This weakness will limit the ability to support complex machine functionality and in manufacturing of complex geometry. Even simple geometry will require a long command lines to execute.

In STEP-NC, Geometric information is defined by machining features with machining operations termed "Workingsteps" performed on one or more features. These Workingsteps provide the basis of a "Workplan" to manufacture the component [7]. Simple geometries can be executed with reasonable and simple command lines, the STEP-NC codes in the example, shown user about drawing or making a polyline, which is a connected series of one or more line segments. A polyline is created by specifying the end points of each segment. The entity CARTESIAN_POINT will furnish the codes with the coordinates of the polyline segments. STEP-NC does provide users with features that support geometric information. Fig. **10** shows an example of STEP-NC part program for polyline, which is simple and easily understandable.

```
#170= POLYLINE('SLOT2',(#171,#172));

#171= CARTESIAN_POINT('SLOT2',(0.,0.,0.));

#172= CARTESIAN_POINT('SLOT2',(25.,0.,0.));
```

Figure 10: An Example of STEP-NC Part Program.

Difficult Changes *vs.* Manageable Changes

Traditional NC data (ISO 6983) contain technical information implicitly. Because of this, it is impossible to manage, modify, and verify NC data easily and avoid human error [21]. In STEP-NC, modification at the shop-floor can be saved and fed back to the design department [7].

A last minute changes and corrections to complex programming of G & M Code are difficult to manage on the shop floor. As a G & M-code part program is made up of a coded set of numbers for axis movements, it is not easy for machine operators to understand the operational flow, machining condition and specification of tools only by reading the low-level part program. In particular, it makes it more difficult, not only in finding which part happens to cause problems, but also modifying the program for solving these problems. A simple change in a complex G & M code program such as a change in diameter for round hole, would be difficult to handle as G & M code execute the commands in machine instructions little by little. For example, for the process of drilling a hole with a depth of 20 mm, the G & M code will drill with an ascending depth of 5 mm until it reaches 20 mm. Any last minute changes for the change in depth will be tedious to handle as the depth drilling process of 5 mm will be in a blocks of codes instead of line of code, and traceability of where changes should be done is difficult.

For STEP-NC codes, last minutes changes at the shop floor is much more manageable, with STEP-NC features as being more user friendly than machine friendly.

Limited Axis *vs.* Dependable Axis

Current technology has allowed, CNC machines to receive more than five axis machining data required to manufacture a part. In G & M Code, geometric data is referred to as machine control data (or MCD), MCD provides a very low level of instruction: tool, axes positions, feed and speed. The problem with MCD programs is that they are not portable or adaptable. Portability is a problem since unique axes position data must be generated for each machine control combination, on which the part is to be run. Adaptability is a problem because no information is provided to the machine to help it adapt to real-time changes in machining dynamics or machine tool alignment [22].

STEP-NC code can support the needs for more than five axis milling. STEP-NC codes are based on tool centre programming rather than machine movement. In tool centre programming, cutter movement data, instead of axis movement data, is sent to CNC machines.

STEP-NC allows tool centre programming to define program geometry as cutter movement data, instead of axes movement data. STEP-NC also provides rich, high level information about the part features, materials, cutters, and dimensional tolerances. Tool centre programming is similar to robotic 6D pose representation. Motion is defined as a 3D tool-tip position (X,Y,Z) and a 3D tool axis orientation (I,J,K). For each tool centre programming (TCP), (X,Y,Z,I,J,K), the CNC controls the two rotation axes so that the tool is positioned and oriented as specified. In addition, the CNC controller performs tool offset compensation along the tool axis according to the position of the tool tip in the proper position and orientation [22].

Primitive *vs.* Modern (Intact)

In G & M code, information is reduced to primitive instructions. For example, the simple steps or command of drilling a hole with a depth of 60 mm will have a long commands line which is represented in machine language. As what shown in Fig. **11**, the drilling of depth 60 mm was done ascendingly by 30 mm and the instructions are being segregated to more than one blocks of commands.

In STEP-NC, information remains intact in its context and not reduced to primitive instructions, the direction or commands given by the program is high level enough to be trace and understand by shop floor programmers or users, as shown in Fig. **12**.

Figure 11: G & M Code Part Program.

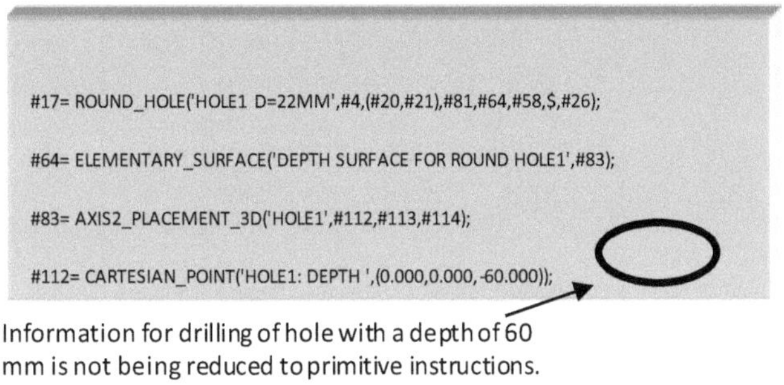

```
#17= ROUND_HOLE('HOLE1 D=22MM',#4,(#20,#21),#81,#64,#58,$,#26);

#64= ELEMENTARY_SURFACE('DEPTH SURFACE FOR ROUND HOLE1',#83);

#83= AXIS2_PLACEMENT_3D('HOLE1',#112,#113,#114);

#112= CARTESIAN_POINT('HOLE1: DEPTH ',(0.000,0.000,-60.000));
```

Information for drilling of hole with a depth of 60 mm is not being reduced to primitive instructions.

Figure 12: STEP-NC Code Part Program.

Implicit *vs.* Explicit

The implicit technical information in ISO 6983, reflects the machining knowledge of experienced NC programmers and machinist, but disables others to obtain machining knowledge directly from the NC data. The technical information is contained implicitly inside the G & M Code. As a result, it is impossible to manage, modify and verify the NC data easily and avoid human error.

In order to obtain technical information, NC data are analyzed based on a next-generation CNC control language (ISO 14649) [23], which contains technical information explicitly. The new ISO 14649 data model includes rich information such as machining features, machining operation, and machining conditions enabling feature-based programming [24]. STEP-NC provides a complete and structured data model, linked with geometric and technological information, so that no information is lost between the different stages of the process [19].

Unidirectional *vs.* Bidirectional

One of the weakness of G & M Code is that there is no bi-direction exchange of information between CAD/CAM systems and CNC machine tools. As a consequences, changes in shop floor will result in rework in planning.

For example, if there is variation in clamping and should be considered in setup of CNC machine tool, it must be calculated once again in CAD/CAM system. Another example is that collision avoidance depends on the geometries of work piece, the machine tools, and the used tools in each machine tool. To avoid the collision if there is a change in machine tool, the determination of cutting speed, feed rate, and depth of cut must be performed again in CAD/CAM system for the machine tool [25]. This weakness will reduce the flexibility, safety and the responsiveness of the whole manufacturing process.

In G & M code, the standard assumes information flow is from CAD to the shop floor, and doesn't enable feedback of experience from the shop-floor to the designer [12]. It only supports one-way information flow from design to manufacturing; the changes on the shop-floor cannot be directly feed back to the designer. Hence, important experiences on the shop-floor can hardly be preserved [19].

STEP-NC supports a bi-directional information transfer between CAD/CAM and CNC. As a result, modifications for the information about machining tasks and technological data on the shop-floor can be saved and transferred back to the planning department [19].

Unintuitive *vs.* Intuitive

In G & M code, commands are written in single alphabets and two digit numbers. It makes the programming code unintuitive or not easily graspable by users intuition.

A G-code based part program only contains low-level information that can be described as "how-to-do" information. The CNC machine tools, no matter how capable they are, can do nothing but "faithfully" follow the G-code program. It is impossible to perform intelligent control nor machining optimization [7].

```
#7=WORKPIECE('PART 2',#13,0.01,$,$,$,(#9,#10,#11,#12));

#9=CARTESIAN_POINT('CLAMPING_POSITION1',(25.,25.,-20.));

#10=CARTESIAN_POINT('CLAMPING_POSITION2',(205.,25.,-20.));

#11=CARTESIAN_POINT('CLAMPING_POSITION3',(25.,155.,-20.));

#12=CARTESIAN_POINT('CLAMPING_POSITION4',(205.,155.,-20.));
```

Figure 13: STEP-NC Part Program.

In STEP-NC, commands are interpretable based on instruction needed which makes the programming code intuitive. With reference to Fig. **13**, machine operators or users can easily understand that for the making of Workpiece 'PART 2' it will have 4 clamping positions where each position will be clamped based on the Cartesian point provided in the codes.

Unstandardized *vs.* Standardized

Due to the lack of further development of ISO 6983 or G & M Code, the immense variety of tool configurations, and little demand or interoperability, few machine controllers (CNCs) adhere to this standard. Extensions and variations have been added independently by manufacturers, and operators of a specific controller must be aware of the differences between each manufacturers' product. Today, sample main manufacturers of CNC systems are GE Fanuc Automation, Siemens and Mazak, but there still exist many smaller and/or older controller systems.

The unstandardized programming of G & M Code has made the exchanged of information between departments to be difficult. Updates and reusability of the command lines are also hard to be done as update commands of G & M Code are different among NC machine companies.

In G & M code, CNC programs are not exchangeable [19]. But with the STEP-NC, due to standardization by the ISO committee, post-processors will be eliminated as the interface does not require a machine-specific information. Machine tools are safer and more adaptable because STEP-NC is independent from the machine tool vendor [19].

STEP-NC allows multi-direction data exchange among different system and among controllers of machine tools. The multi-direction data exchange will makes the manufacturing process to be safer, more flexible and more responsive.

By using STEP-NC, data exchange among different type of data format is not a problem anymore because it uses STEP as a standard data format. The use of standard data format will bring the flow of information in supply chain especially from design to procurement and to production faster and more reliable [25].

STEP-NC makes manufacturing set-up faster and more flexible because there is no need to specify specific machine tools in CAD/CAM systems.

SUMMARY

According to Xu [13], providing feedback from manufacturing to design, process planning, and quality control promises improved products and significant cost and time savings. G-code format data are useful passed forward to machining but are useless for feedback (except for changed G-code). The STEP-NC data format is better for providing feedback, in addition, STEP integration might be extended to include feeding back production run data such as actual tool selection and cutting time by modeling these data using established STEP techniques. If these data were available in process planning, they would enable long-term optimization and better NC visualization and verification. If the data were available in design, they would facilitate improving design-for-manufacturing techniques and determining the consequences of design choices on manufacturing cost [13]. Finally, it should be mentioned that this work is a continued research based on the authors' previous effort as published in [26].

ACKNOWLEDGEMENT

Declared none.

CONFLICT OF INTEREST

The author(s) confirm that this chapter content has no conflict of interest.

REFERENCES

[1]   A. Kumar and J. Saha, "Automatic Data Extraction From ISO10303-21 (STEP) for Feature Recognition," *Arab Res. Inst. Sci. Eng.,* vol. Vol. 4, pp. 129-136, 30 June 2008.

[2]   D. Gibbs, *An Introduction to CNC Machining* Second Edition ed.: Cassel Publishers Ltd. Great Britain, 1987.

[3]   ISO, "ISO 6983-1-Numerical control of machines-program format and definition of address words-Part 1: data format for positioning, Line motion and contouring control systems.," ed, 1982.

[4]   X. W. Xu, "Realization of STEP-NC enabled machining," *Robot. Comput.-Integr. Manuf.,* vol. 22, pp. 144-153, 2006.

[5]   S. T. Newman, Nassehi, A., Xu, X.W., Rosso Jr., R.S.U., Wang, L., Yusof, Y., Ali, L., Liu, R., Zheng, L.Y., Kumar, S., Vichare, P., Dhokia, V., "Strategic Advantages of Interoperability for Global Manufacturing Using CNC Technology," *Robot. Comput.-Integr. Manuf.,* vol. 24, pp. 699– 708, 2008.

[6]   ISO, "ISO 14649 Part 10 General Process Data," ed, 2002.

[7]   X. W. Xu and S. T. Newman, "Making CNC Machine Tools More Open, Interoperable and Intelligent - A Review of the Technologies," *Comput. Ind.,* vol. 57, pp. 141-152, 2006.

[8]   M. F. Hushim, *et al.,* "Comparative Study Between Physical File Part-21 STEP-NC and G&M Code ISO 6983," in *Proceedings of the International Conference of Application and Design in Mechanical Engineering (ICADME 2009),* 2009.

[9]   Steptools. (2009, June 16). *STEP Application Protocols.* Available: http://www.steptools.com/library/standard/step_2.html

[10]   SC4ONLINE. (2010, 16 June 2010). *STEP Overview.* Available: http://www.tc184-sc4.org/SC4_Open/SC4%20Legacy%20Products%20%282001-08%29/STEP_%2810303%29/

[11]   S. B. Brunnermeier and S. A. Martin, "Interoperability Cost Analysis of the U.S. Automotive Supply Chain," Research Triangle Institute, March 1999.

[12]   W. Maeder, *et al.,* "Standardisation of the Manufacturing Process: The IMS STEp-NC Project.," presented at the National Network of Competence on Integrated Production and Logistics, Saas Fee, Switzerland, 2002.

[13]   X. W. Xu, *et al.,* "STEP-compliant NC research: the search for intelligent CAD/CAPP/CAM/CNC integration," *Int. J. Prod. Res.,* vol. Vol. 43, pp. 3703–3743, 1 September 2005 2005.

[14]   S. H. Suh, *et al., Theory and Design of CNC Systems*: UK:Springer, 2008.

[15]   S.-N. Newsletter, "STEP-NC Newsletter," *STEP-NC Newsletter,* November 2000 2000.

[16]   ISO, "ISO 14649-11, Data Model for Computerized Numerical Controllers. Part 11 Process Data for Milling.," ed, 2004.

[17]   Wikipedia. (2010, July, 16 June 2010). *Low-level programming language.* Available: http://en.wikipedia.org/wiki/Low-level_programming_language

[18]   E. Fortin, *et al.,* "An Innovative Software Architecture to Improve Information Flow from CAM to CNC," *Comput. Ind. Eng.,* vol. 46, pp. 655-667, 2004.

[19]   Z. Xioming, *et al.,* "A 3-D Simulation System for Milling Machining Based on STEP-NC," In: *Intelligent Control and Automation, The Sixth World Congress on WCICA,* Dalian, China, 2006.

[20]   X. L. Qiu, *et al.,* "Feature-Based Process Planning using STEP-NC," In: *The International Technology and Innovation Conference,* Hangzhou, China, 2006.

[21]   H. Yamada, *et al.,* "Analysis of ISO 6983 NC Data Based on ISO 14649 CNC Data Model," In: The SICE-ICASE International Joint Conference, Bexco, Busan, Korea, 2006.

[22]   S. Ventakesh, *et al.,* "Validating Portability of STEP-NC Tool Center Programming," In: Proceedings of IDETC/CIE, ASME International 25th Computers and Information in Engineering Conference, Long Beach CA, USA, 2005.

[23]   ISO, "ISO 14649-1:2003 Industrial Automation Systems and Integration - Physical Device Control - Data Model for Computerized Numerical Controllers - Part 1: Overview and Fundamental Principles," ed, 2003.

[24]   S. H. Suk, *et al.,* "On the Architecture of Intelligent STEP-Compliant CNC," *Int. J. Comput. Integr. Manuf.,* vol. 15, pp. 168-77, 2002.

[25]   J. Suteja, "The Role of STEP-NC in Improving the Performance of Supply Chain," *Jurnal Tek. Ind.,* vol. Vol. 7, pp. 113-118, December 2005.

[26]   Yusri Yusof, Noor Diana Kassim & Nurul Zakiah Zamri Tan, The development of a new STEP-NC code generator (GEN-MILL), *Int. J. Comput. Integr. Manuf.,* vol. 24, Issue 2, pp. 126-134, 2011

Send Orders for Reprints to reprints@benthamscience.net

CHAPTER 4

Characterization of Reconfigurable Stewart Platform for Contour Generation and Active Vibration Isolation Applications

Nagarajan Thirumalaiswamy[1,*] and Kumar G. Satheesh[2]

[1]Department of Mechanical Engineering, Universiti Teknologi Petronas (UTP), 31750 Tronoh, Perak, Malaysia and [2]Precision Engineering and Instrumentation Laboratory, Department of Mechanical Engineering, Indian Institute of Technology Madras, India

Abstract: A parallel manipulator is a closed-loop mechanism in which the end-effector is connected to the base by at least two independent kinematic loops. A general description of these types of manipulators is explained with examples and applications. With multiple closed loops, stiffness of the manipulator is typically improved because the multiple leg connectors sustain the payload in a distributive manner. Off-late the re-configurability of the platforms gain more research interest among researchers for its increasing practical applications in industries. The commercial hexapods that are available in the market are mission specific with no choice offered between structural rigidity and dexterity to use the same platform for other applications.

In this chapter, an effort is made to characterize the parameters for developing a reconfigurable Stewart platform for contour generation and vibration isolation applications. The limited treatment of the platform characteristics leads to the lack of an efficient methodology for determining the optimum geometry for this task. A solution is provided through the formulation of dimensionless parameters in combination with a study on the generic parameters like configuration. The variable geometry approach for the reconfiguration of Stewart platforms is adopted in detail for four different platforms, and a generic approach is formulated after studying different parameters.

A stiffness model developed for contour generation application is used in tandem with this generic approach to identify the trajectory with maximum stiffness for complex contours. An extensive study on the effect of the identified parameters on the performance and characteristics of Stewart platform for both the applications are performed. Simulations, in order to study different contours are performed to obtain the trajectory with maximum stiffness. The configuration best suited for contour generation also is identified and the data set required to move the tool for a specific trajectory is also identified. The effect of the identified dimensionless parameter on the system

---

*Address correspondence to Nagarajan Thirumalaiswamy: Department of Mechanical Engineering, Universiti Teknologi Petronas (UTP), 31750 Tronoh, Perak, Malaysia; Tel: + 6053687028; Fax: + 6053656461; Email: nagarajan_t@petronas.com

Dan Zhang and Zhen Gao (Eds.)

performance is also studied. The parameters are tested for vibration isolation application also and the results are used in designing the test rig.

The Stewart platform test rig developed is fed with the dataset obtained for different trajectories and error analysis is performed to validate the simulation results. An algorithm is developed on the basis of kinematic path control for conducting the experiments to find the tracking performance. The results are used to establish the importance of dimensionless parameters for reconfiguration of Stewart platform. It is proposed that this methodology could be adopted for any application to develop a complete set of design tool for any new reconfigurable Stewart platform. Experimentations to identify the natural frequency of the developed Stewart platform were performed to ascertain the frequency used for simulation studies. A novel concept of Multilevel reconfigurable Stewart platform is introduced to overcome difficulties and also for effective performance in both the applications.

Keywords: Stewart platform, reconfigurable, parallel manipulator, robot, characterization, contour generation, trajectory, vibration isolation, cubic configuration, hexapod, joint vector, maximum stiffness.

1. INTRODUCTION

As science and technology of robotics originated with the spirit of developing mechanical systems to carry out tasks normally ascribed to human beings, it is quite natural to use open loop serial links and joints as robot manipulators. Such robot manipulators have the advantage of sweeping workspaces and dexterous maneuverability like the human arm, but their load capacity is rather poor due to the cantilever structure. Hence, for applications where high load carrying capacity, good dynamic performance and precise positioning are of paramount importance, it is desirable to have an alternative to conventional serial manipulators. One looks to the biological world for possible solutions and observe that limbs or fingers are actuated in parallel for stability, precise work and handling heavy loads in animals. This research is a continued work based on the authors' previous effort as published in [1].

1.1. Serial *vs.* Parallel Manipulators

Hence, for applications where high load carrying capacity, good dynamic performance and precise positioning are of paramount importance, it is desirable to have an alternative to conventional serial manipulators. One looks to the biological world for possible solutions and observe that:

1. the bodies of load-carrying animals are more stably supported on multiple in-parallel legs compared to the biped human

2. human being also use both the arms in cooperation to handle heavy loads and

3. for precise work like writing , three fingers actuated in parallel are used.

Upon extrapolation it can be expected that robot manipulators, having the end-effector connected to the ground *via* several chains and having actuations in parallel, to be attractive for certain applications. The last two decades have witnessed considerable research interest in this direction. Apart from classifying robot manipulators into serial and parallel types, classification based on the open-loop and closed nature is also in practice. However, it should be borne in mind that the two classifications are not identical. Though open-loop manipulators are always serial, and parallel ones are always with closed loop(s), it is possible to have closed-loop manipulators which are serial in nature. Traditionally, most manipulators are serial structures *i.e.* they are made up of rigid body links that are connected one after another by simple joints (usually revolute or prismatic). The human arm is a good example of a serial manipulator. Most common industrial manipulators have seven links (counting ground) inter connected by six simple joints.

These structures are well constructed, highly developed, and are widely used in industrial applications. Serial manipulators do not have closed kinematical loops, and actuated at each joint along the serial linkage. Accordingly, the weight of the actuators that are located at each joint along the serial linkage can account for a significant portion of the load carrying capability of the arm.

Compared to a serial manipulator, a parallel manipulator is a closed-loop mechanism in which the end-effector (mobile platform) is connected to the base by at least two independent kinematic loops. The direct position kinematics of serial manipulators is straightforward while the inverse kinematics is quite complicated requiring solution to a system of nonlinear equations. With the multiple closed loops, the stiffness of the manipulator is typically improved because the multiple leg connectors sustain the payload in a distributive manner as long as the mechanism is far from a singular configuration. In addition, end-point positioning error is also reduced when compared to a serial parallel manipulator due to non accumulation of errors.

1.2. Stewart Platform

The Stewart platform, proposed by D. Stewart, as an aircraft simulator is a six DOF parallel manipulator where the end-effector is attached to a movable plate supported in parallel by six linear actuated links. In the last several years, there has been an increase in the applications of the Stewart platform in developing robotic devices. As a manufacturing manipulator it has two fundamental characteristics that set it apart from machine tools and industrial robots [2]. It is a closed kinematic system with parallel links. The links take the load axially, making the manipulator far more rigid in proportion to size and weight than any serial link robot.

1.3. Applications of Stewart Platform

1.3.1. Vibration Isolation

Vibrations propagating into the mechanical systems can cause many problems at different levels causing performance degradation for sensitive systems of both ground and space applications. Vibration isolation is defined as the attenuation of the response of the system, to cut-off all the disturbances after corner frequency and allowing all the signals below it to pass faithfully. A multi-purpose generic active isolator based on Stewart platform with voice coil actuators developed by Jet Propulsion Laboratory.

1.3.2. Semiconductor Technology

The semiconductor industries have used the robots to handle wafers and also in wafer alignment which is translated into major cost savings. Application of clean room robots extend to the areas of machine loading, machine unloading, parts transfer, assembly, packaging and testing processes.

1.3.3. Astronomy Field

Two types of robots are used in this field of research: Manipulators and Rovers. Manipulators are used for holding the telescoping equipments and for antenna positioning. Rovers are set out on the planetary surfaces to perform certain human-like activities.

1.3.4. Telecommunication

Micro systems technology and alignment of fibers are the important areas where robots play a useful role. By the application of network robotics, the robot can become an information terminal with a user-friendly network-human interface. Telesurgery is another important application which talks about the successful interaction of man and robots.

2. MOTIVATION

The important areas where Stewart platform has solved the pertinent time immemorial problems are Machine tool applications in manufacturing industries, precision pointing and Vibration isolation for ground and space applications. The systems that are available in the market, the commercial hexapods, are mission specific with no choice between structural rigidity and dexterity to use the same platform for other applications [3].

This work presents the initial investigations with the above mentioned applications as the primary objective. The proposed methodology is expected to provide a holistic approach to develop a complete set of design tool for any new reconfigurable Stewart platform.

3. METHODOLOGY

Cubic configuration was proposed for vibration control applications and so it is expected for 3-3 configuration to have better performance than the 6-6 configuration of similar geometrical parameters [4]. In a comparison study of workspace analysis of two reconfigurable Parallel Machine Tools, inverse kinematics solution was provided including factors like radius of mobile platform, strut length and spherical joint setting angle by Lin.

Other important implications, which make it possible to be used for other applications, should be studied. Such a capability is far superior to the trial and error procedure in situation where task requirements vary.

So it is desirable to characterize parameters to develop a reconfigurable Stewart platform for contour generation and vibration isolation applications by variable geometry approach. An algorithm is also developed for stiffness analysis along the variable geometry approach and the stiffness is to be studied for different contours. Apart from that, the effect of the identified parameters on the performance and characteristics of Stewart platform for both the applications are studied.

4. CHARACTERIZATION FOR CONTOUR GENERATION

An explicit study on λ and γ is avoided since they are directly dependent on $\mathbf{b}_i$ (referred to as joint vector). Fig. **1** represents the PeilPOD3-3 developed for this research work and $\mathbf{s}_i$ is unit vector pointing from A_i to B_i.

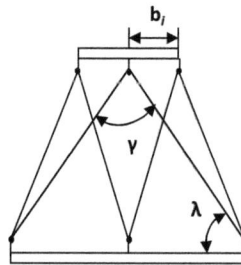

Figure 1: Modular parameters.

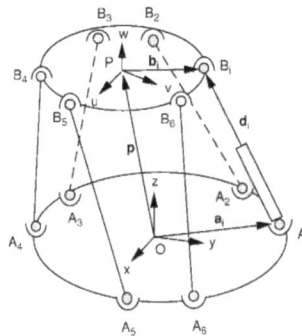

Figure 2: General Stewart platform.

4.1. Degree of Freedom

It is the number of independent variables (or coordinates) required to completely specify the configuration of the system. By Grübler's criterion the degree of freedom of a system could be obtained from the following expression:

$$F = \lambda(n - j - 1) + \sum_i f_i - f_p$$

where, F = degree of freedom, λ - 3 or 6 for planar or spatial mechanism, n – number of links, j – number of joints, f_i – degree of freedom of individual joints and f_p – passive degree of freedom.

Note: Through passive degree of freedom there would be no transfer of motion of torque.

To calculate DOF for the Stewart platform given in Fig. **2**.

The parameters are:

$n = 14$ (2 for each leg + top & bottom platform)

$j = 18$ (12 spherical + 6 prismatic)

$f_i = 1$ for prismatic joints

$= 3$ for spherical joints

$f_p = 6$ (along the axis of each leg)

$\lambda = 6$ (spatial mechanism)

Substituting in the Grübler's criterion:

$$F = \lambda(n - j - 1) + \sum_i f_i - f_p$$

$$= 6\,[14 - (12 + 6) - 1] + [(12 \times 3) + (6 \times 1)] - 6$$

$$= 6 \text{ DOF}$$

4.2. Position Analysis of a Stewart Platform

For inverse kinematics, the position vector **p** and rotation matrix $^{A}R_B$ of frame B with respect to A are given (Fig. **2**) and the limb lengths are found to be [5],

$$d_i = \pm\sqrt{[\mathbf{p} + {^A}R_B{^B}\mathbf{b}_i - \mathbf{a}_i]^T[\mathbf{p} + {^A}R_B{^B}\mathbf{b}_i - \mathbf{a}_i]} \tag{1}$$

4.3. Jacobian Analysis of Parallel Manipulators

Gosselin and Angles studied the overall Jacobian matrix J is given by [6],

$$J = J_q^{-1} * J_x \begin{pmatrix} s_1{^T} & (b_1 \times s_1)^T \\ s_2{^T} & (b_2 \times s_2)^T \\ s_3{^T} & (b_3 \times s_3)^T \\ s_4{^T} & (b_4 \times s_4)^T \\ s_5{^T} & (b_5 \times s_5)^T \\ s_6{^T} & (b_6 \times s_6)^T \end{pmatrix} = \tag{2}$$

4.4. Simulation Results and Discussion

The topview of 6-6 configuration is shown in Fig. **3(a)** and the corresponding side view is shown in Fig. **2**. It is common for both Tsai6-6 and PeilPOD6-6 platforms. Similarly Fig. **3(b)** shows the topview of 3-3 configuration and Fig. **1** gives the sideview.

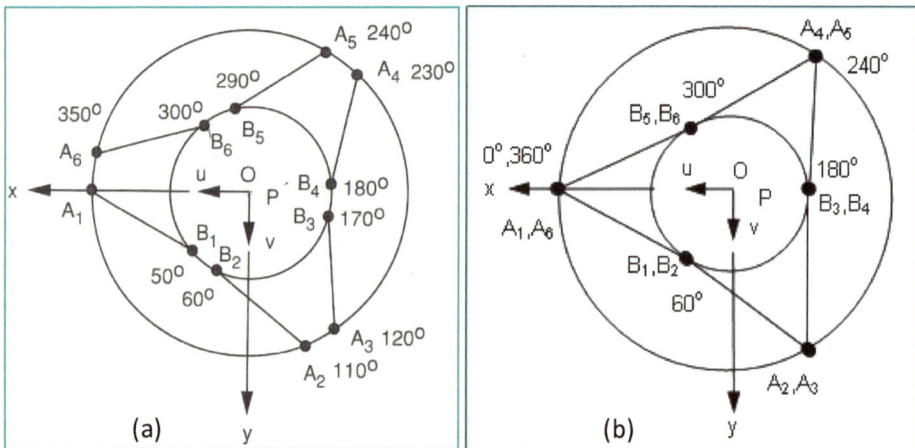

Figure 3: (a) & (b) Top views 6-6 and 3-3 configuration.

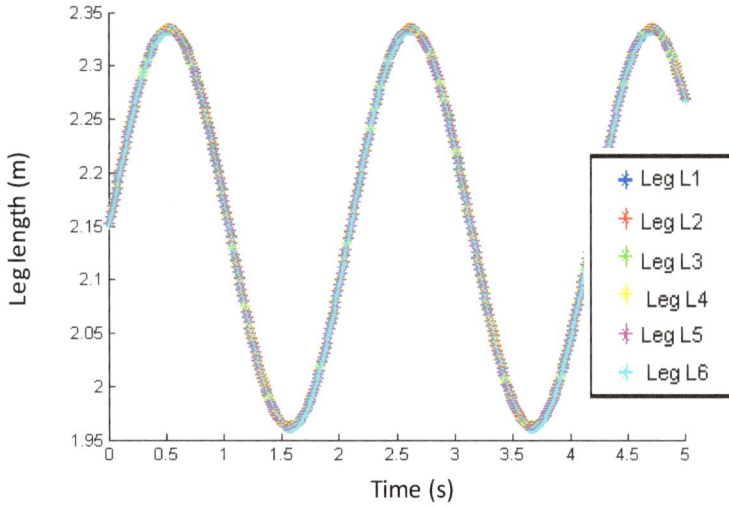

Figure 4: Leg lengths for Sine wave.

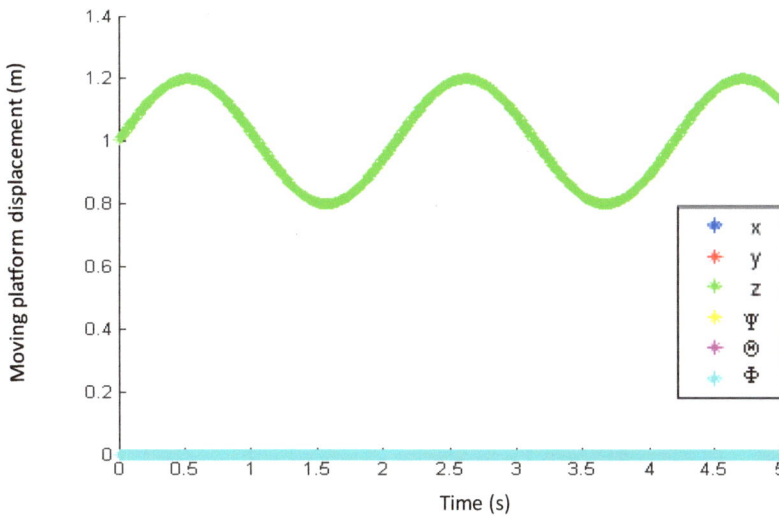

Figure 5: Platform displacement for Sine wave.

4.5. Validation of Algorithm

Sinusoidal and circular trajectories are chosen for the purpose of validation. Eqns. (3) and (4) provide the displacements of centroid, for those trajectories respectively.

$$\Delta x = [0, 0, (1+0.2*\sin(w*t)), 0, 0, 0]^{\mathrm{T}} \tag{3}$$

$$\Delta x = [p_x + r * \sin \beta, p_y + r * (1 - \cos \beta), p_z, \psi, \theta, \phi] \tag{4}$$

Fig. **4** shows the leg length variations for sinusoidal trajectory of Tsai 6-6 platform. The displacements of all the legs are equal and the overlapping curves obtained are sinusoidal. The z-axis disturbance obtained in the moving platform is observed in Fig. **5**.

4.6. Influence of Joint Vector

To study the effect of joint vector simulation are done only for the 3-3 configurations of Tsai and PeilPOD. For different position vectors of the joints at the moving platform simulations are performed and the maximum stiffness obtained is recorded. The stiffness shows a gradual increase with decrease in η (*i.e.* increase in the moving platform diameter) for both Tsai 3-3 platform and PeilPOD platforms. For Tsai platform it increases upto η = 0.4 and starts to fall down. For PeilPOD stiffness increases upto η = 0.7. Any increase in η beyond these values does not show any improvement. Fig. **6** shows the semi-logarithmic plot of the observed stiffness for both platforms. This chart would serve as a design tool to choose between the stiffness and the dimensions of the top and bottom platforms.

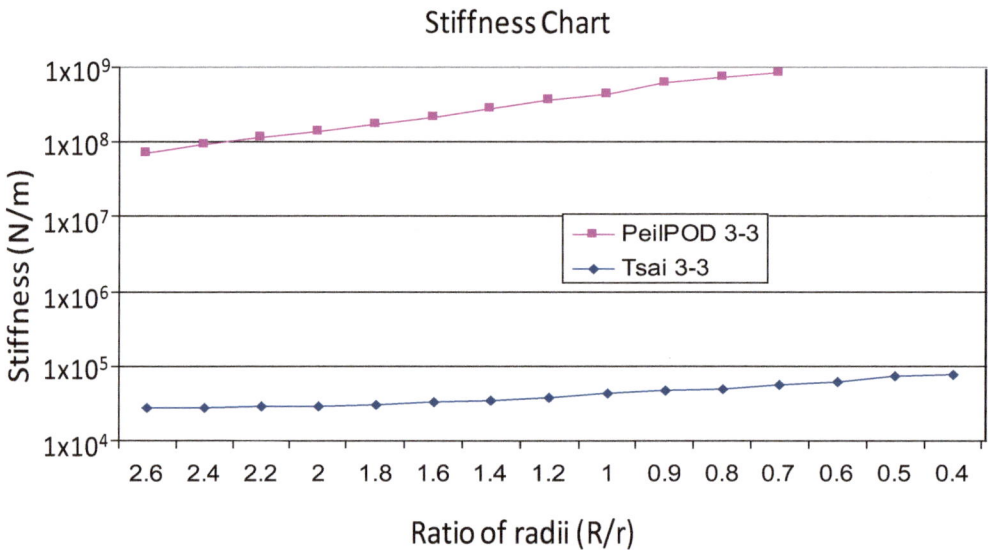

Figure 6: Stiffness variation for Spiral trajectory.

5. CHARACTERIZATION FOR ACTIVE VIBRATION ISOLATION

From the Newton's equation of motion and the Euler's equation for the platform the complete dynamic equations of the platform in closed form is obtained as,

$$\mathbf{J}\begin{bmatrix} \ddot{\mathbf{t}} \\ \alpha \end{bmatrix} + \mathbf{J}_d\ddot{\Phi} + \sigma = \mathbf{HF} + \begin{bmatrix} r\,\mathbf{F}_{ext} \\ r\,\mathbf{M}_{ext} \end{bmatrix} \tag{5}$$

Equation (5) provides the task space dynamics equations where $\mathbf{F}_{ext}$ and $\mathbf{M}_{ext}$ are the external force and external moment to be controlled. For the purpose of simulation these parameters are taken as the source of disturbance for vibration isolation.

5.1. Control Law

To generate counter vibratory forces a PD control algorithm is used, for task-space. It is chosen based on the simulation results obtained without control (Fig. 7). The proportional constants were decided through simulation.

$$\mathbf{F}_{task} = \mathbf{diag}[K_{p1}\ K_{p2}\ K_{p3}](\mathbf{t}_0 - \mathbf{t}) + \mathbf{diag}[K_{v1}\ K_{v2}\ K_{v3}](\dot{\mathbf{t}}_0 - \dot{\mathbf{t}}) \tag{6}$$

$$\mathbf{M}_{task} = \mathbf{diag}[K_{p4}\ K_{p5}\ K_{p6}](\Theta_0 - \Theta) + \mathbf{diag}[K_{v4}\ K_{v5}\ K_{v6}](\omega_0 - \omega) \tag{7}$$

5.2. Influence of Configuration

3-3 and 6-6 configurations are studied in the time domain. Tsai3-3 and Tsai6-6 platforms are simulated for the sinusoidal input of 70 Hz (refer section 6.1) and amplitude 300 N. The initial conditions taken for the simulation are $\mathbf{t}_0 = [0.1, 0.0, 0.3]^T$ m and $\Theta_0 = [0.0, 0.0, -0.2]^T$ m. Fig. **7** provides the comparison of linear displacement of Tsai 3-3 and Tsai 6-6 platforms without control in z direction and comparison with control is provided in Fig. **8**. It is observed that 3-3 configuration has a settling time of 10ms lesser than the 6-6 configuration for the z direction. Since the disturbance produced is in the z-direction alone displacement obtained in other directions is lesser than that obtained in the z-direction.

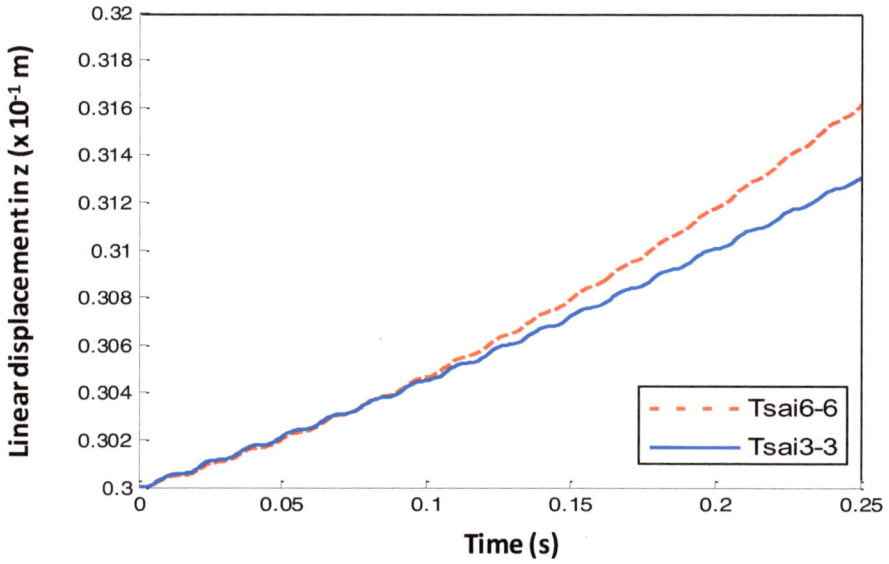

Figure 7: Linear displacement of the platform in z-direction (without control).

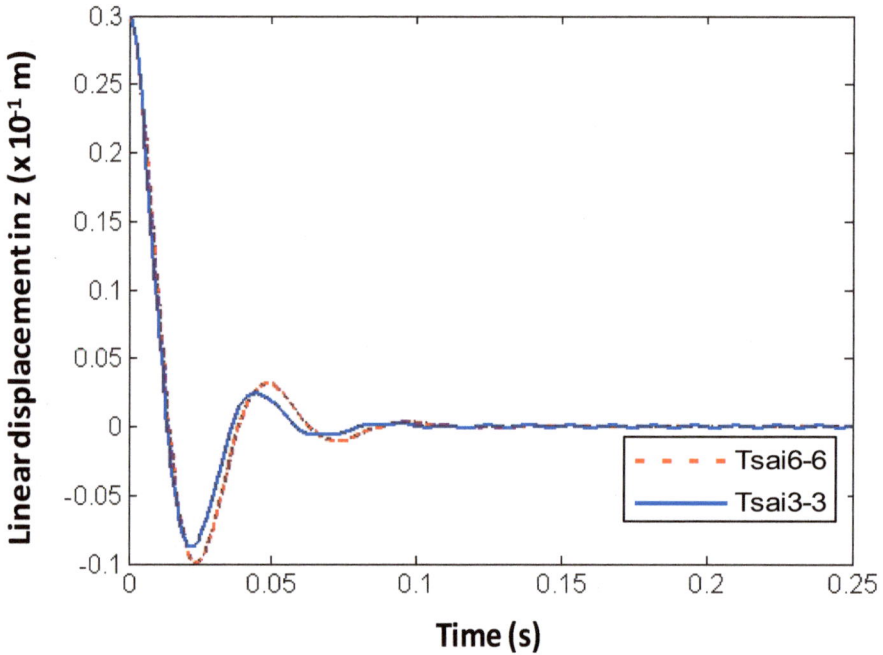

Figure 8: Linear displacement of the platform in z-direction (with control).

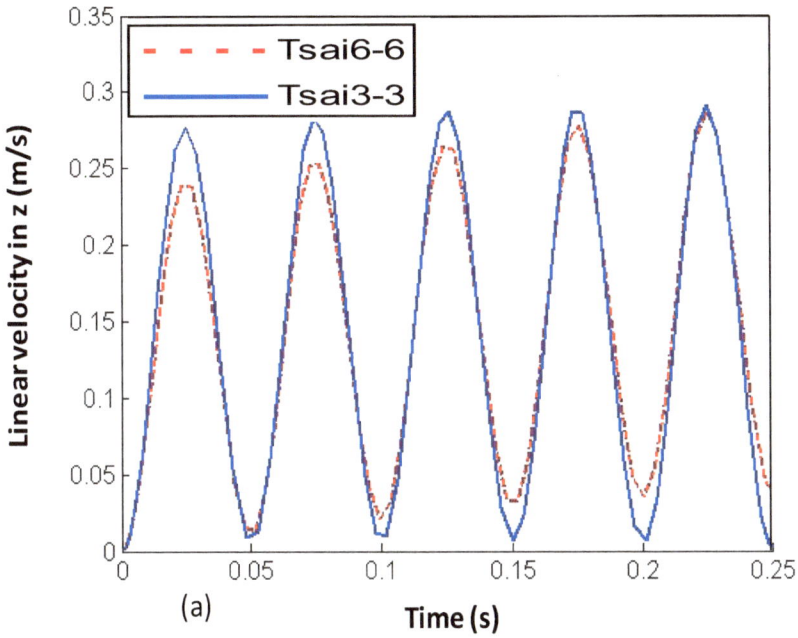

Figure 9: Linear velocity of the platform in z-direction (without control).

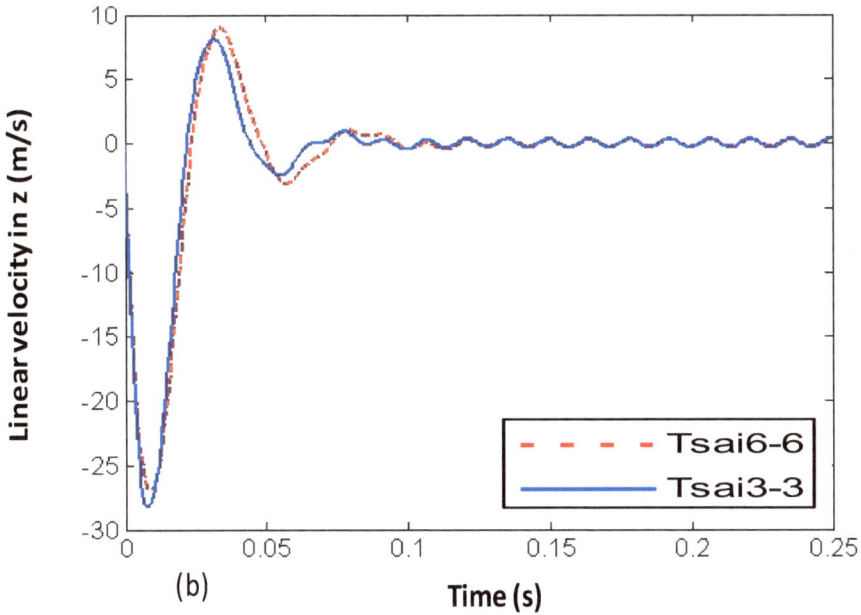

Figure 10: Linear velocity of the platform in z-direction (with control).

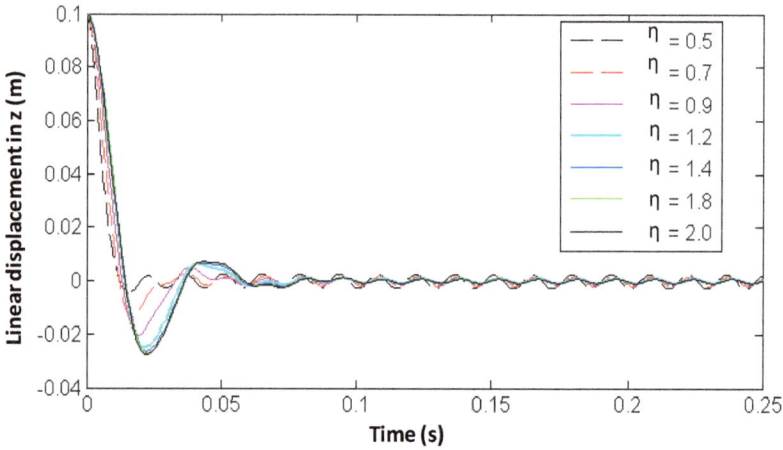

Figure 11: Linear displacement of the platform in z-direction.

PD control gives better performance for the 3-3 configurations. The overshoot in each case is considerably reduced for 3-3 configuration. The linear velocities of centroid of the moving platform in z-direction are presented in Figs. **9** & **10**. The output of simulation shows a better performance for the 3-3 configuration over 6-6 configuration for active vibration isolation applications. The same trend is observed with other platforms like PeilPOD3-3 and PeilPOD6-6 too.

5.3. Influence of Joint Vector

As in the case of stiffness study, the joint vector is simulated only for the 3-3 configurations of Tsai and PeilPOD, and results are presented for Tsai platform. The dimensionless parameter η, the ratio R/r, is varied from 0.3 to 3.0. The effective active range of η is observed to be 0.5 to 2.0. An increase or decrease beyond this range introduces a highly unstable behavior into the system which is beyond the control of the applied PD control law. Fig. **11** provides the plot for linear displacement in z-direction for different η. As η is increased from 0.5 to 2.0 the settling time is increased. The overshoot also reduces with reduction in η.

6. MULTILEVEL STEWART PLATFORM (MSP)

A novel concept of modular reconfigurability is introduced which could be an innovative practical solution for the problems mentioned before. It is called

Multilevel Stewart platform because of its multilevel structure, as shown in Fig. **12** this active structure has one Stewart platform mounted on the other.

The proposed platform is shown in Fig. **12(a)** which is a variation of the one shown in Fig. **12(b)**. Structures as the one shown in Fig. **12(b)** have been proposed for segregating orientation and positioning task for precision pointing applications and are called hybrid manipulators [5]. Hybrid manipulators posses the advantages of both serial and parallel manipulators from rigidity and workspace point of view.

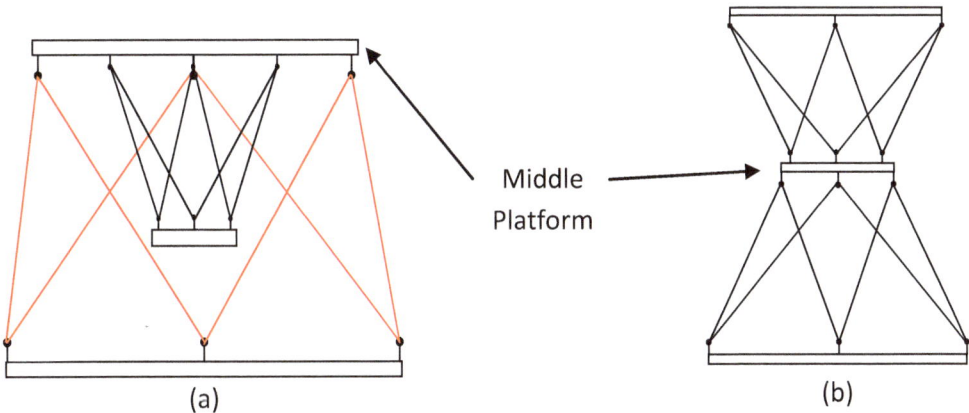

Middle
Platform

(a)

(b)

Figure 12: Multilevel Stewart Platform.

Structures similar to Fig. **12(a)** are already in use in the case of Variax HEXAPOD machines, but the outer structure is passive and plays only the role of a supporting structure. With MSP, while the outside platform can be used to isolate vibration, the inside platform can be used for different purposes like positioning the tool mounted. This former is advantageous for machining applications since it offers structural rigidity and the latter is better suited for positioning application. When the actuators at both levels are used simultaneously for positioning applications a larger work volume is obtained. So the dexterity of the device increases as a whole.

By varying newly defined parameter 'naga' 'η', the ratio R/r, of the middle platform, different stiffness could be obtained and also modifies the parameters for the vibration isolation application. In general, the difference in parameters among the Stewart platforms used for vibration isolation application and motion platforms are:

- stroke is in the order of microns

- frequency response performance in the range of kHz

- and the varied force requirements for different applications

As observed it becomes difficult to use a single-level platform for different applications without changing the actuators which could be solved by using the Multilevel Stewart platform. Actuators with varying capabilities can be mounted at different levels. The problem of lower level actuators carrying the higher level ones can be ignored since the weight of the actuators gets distributed and also for the enhanced reconfigurable capability. Much of research could be converged upon this platform to identify the characteristics and performance to provide a pragmatic solution for reconfigurability problem. The variable geometric approach used in this research work is best suited to improvise the capabilities of this platform.

SUMMARY

A general introduction to Stewart platform is made with reference to various applications. For the vibration isolation applications, it is observed that 3-3 configuration has a settling time of 10 ms lesser than the 6-6 configuration for z direction. Since the disturbance produced is in the z-direction alone displacement obtained in other directions is lesser than that obtained in the z-direction. The output of simulation shows a better performance for the 3-3 configuration over 6-6 configuration for active vibration isolation applications in all the directions. The overshoot in each case is considerably reduced for 3-3 configuration.

The effective active range of the dimensionless parameter joint vector η, the ratio R/r, is observed to be 2.0 to 0.5. An increase or decrease beyond this range introduces a highly oscillatory behavior into the system which is beyond the control of the applied PD control law. As η is decreased, the settling time decreases. The overshoot also reduces with reduction in η. A similar performance was observed for the PeilPOD too.

In case if the setting up time is very critical between the applications it is suggested to have 3-3 configuration for reasonable performance for both the applications. Another significant observation that could be made from analysis of both the results is that a range of 0.7-1.5 provides a moderate performance for both the applications.

ACKNOWLEDGEMENT

The first author wishes to acknowledge Universiti Teknologi Petronas, Malaysia for the facilities to write this research article as a chapter and the both authors wish to acknowledge Indian Institute of Technology Madras, India for carrying out this research work.

CONFLICT OF INTEREST

The author(s) confirm that this chapter content has no conflict of interest.

REFERENCES

[1]   Satheesh G. Kumar, T. Nagarajan, Y.G. Srinivasa, "Characterization of Reconfigurable Stewart Platform for Contour Generation," *Robot. Comput. Integrat. Manuf., vol*. 25, no. 4-5, pp. 721-731, August–October 2009

[2]   Ahmed Abu Hanieh, "*Active Isolation and Damping of Vibrations via Stewart Platform,*" Ph.D thesis, Active Structures Laboratory, Department of Mechanical Engineering and Robotics, Universit´e Libre de Bruxelles, 2003.

[3]   Attallah MM, Rashwan O. "Six DOF Hexapod: Challenge of design and innovation," Term Project. Available: http://biotsavart.tripod.com/hexapod.htm. The American University in Cairo Department of Engineering Mechanical Engineering Unit MENG 356- Mechanical Design I.

[4]   Z. Jason Geng and Leonard S. Haynes, "Six-degree-of freedom Active Vibration Isolation using a Stewart platform Mechanism," *J. Robot. Syst.*, 10(5), 725-744, 1993.

[5]   L. W. Tsai, "Robotic Analysis: The Mechanics of Serial and Parallel Manipulators," *John-Wiley & Sons*, New York, 1999.

[6]   Clement Gosselin and Jorge Angeles, "Singularity Analysis of Closed-loop Kinematic Chains," *IEEE Trans. Robot. Automat.,* Vol. 6 (3), 281-290, 1990.

Send Orders for Reprints to reprints@benthamscience.net

CHAPTER 5

Design of Reconfigurable Robotics Machinery for the Intelligent Manufacturing Systems

Ana Djuric[*] and Waguih ElMaraghy[*]

Intelligent Manufacturing Systems (IMS) Centre, University of Windsor, 401 Sunset Ave., Windsor, ON N9B 3P4, Canada

Abstract: An essential part of a Reconfigurable Manufacturing System (RMS) is self-reconfigurable machines. These machines are based on self-reconfigurable mechanisms that can reshape their structures according to the desired changes. The self-reconfigurable mechanisms can be developed using the reconfigurable modeling theory [10]. This theory is used for development of reconfigurable control systems that intelligently unify reconfiguration and manage the interaction of individual machine control systems within the RMS.

In this paper, an example of 2-DOF Global Kinematic Model (2-GKM) generation and a solution is presented to demonstrate the methodology and application of the reconfigurable modeling theory. The model has combinations of either rotational and/or translational types of joints and all possible positive joint directions in 3D, which is unified using predefined reconfigurable parameters. The reconfigurable parameters are used to control the joint's positive directions and its type (rotational and/or translational). For the symbolic calculation of the 2-GKM dynamic equations, the recursive Newton-Euler algorithm is employed using the symbolic algebra package MAPLE 12. The dynamic model is named Global Dynamic Model (2-GDM). The significance of the 2-GDM is that it automatically generates each element of the inertia matrix A, Coriolis torque matrix B, centrifugal torque matrix C, and the gravity torque vector G, using Automatic Separation Method (ASM). Instead of solving the dynamics of different kinematic structures for the 2DOF machines, the 2-GDM can be used to auto-generate the solution by only defining the reconfigurable parameters. Using 2-GDM equations, a simple example of Ttr (Translation and translational/rotational) structure was used to demonstrate the model capability. Its kinematic, dynamic and control platform solutions are generated using previously defined reconfigurable parameters. The results proved the model's validity.

Keywords: Reconfigurable, machinery, modeling, path, analysis, manufacturing system, robot, kinematics, dynamics, control, newton-euler recursive algorithm, simulation.

*Address correspondence to Ana Djuric and Waguih El. Maraghy:** Intelligent Manufacturing Systems (IMS) Centre, University of Windsor, 401 Sunset Ave., Windsor, ON N9B 3P4, Canada; Tel: (313)577-5387; E-mail: ana.djuric2@wayne.edu; adjuric@uwindsor.ca

INTRODUCTION TO RECONFIGURABLE THEORY

Automated model generation and solution for motion planning and re-planning of automated systems will play an important role in Reconfigurable Manufacturing System (RMS). A highly reconfigurable control system that intelligently unifies reconfiguration and manages the interaction of individual robotic control systems within a Reconfigurable Manufacturing System (RMS) is presented. This research was done under the project *Unified Reconfigurable Open Control Architecture*, UROCA [2]. As UROCA is intended for controlling a wide variety of industrial machines, it has the feature of easy reconfiguration from one machine to another as well as from one application to another with the lowest amount of change. The idea of developing unified kinematic structure started by using the power of comparison between different robotic systems. The study showed that almost 90% of industrial robots have the 6 DOF (Degree of Freedom) and all rotational joints. The differences between them are the joints positive directions, which are affected by the twist angle α_i ($i = 1, 2, ..., 6$). The developed reconfigurable parameters K_1, K_2, K_3, K_4, K_5, and K_6, are used to unify all 6R industrial robots into a model named GPF (General Puma-Fanuc) [3, 4].

The seven solutions for the GPF model are: UKMS (Unified Kinematic Modeler and Solver), RGPFJM (Reconfigurable Generic PUMA -Fanuc Jacobian Matrix), RPFSM (Reconfigurable PUMA -Fanuc Singularity Matrix), RRW (Reconfigurable Robot Workspace), RPFDM (Reconfigurable PUMA -Fanuc Dynamic Model), RPFDM+ (Reconfigurable PUMA-Fanuc Dynamic Model Plus) and RCP (Reconfigurable Control Platform) [3-9].

The GPF model and its seven solutions represent the most important modules that UROCA counts on. They represent a forward step towards having a mobile reconfigurable architecture that accepts either new models or new applications without restoring ourselves to rebuilding from scratch whenever software-, hardware-, control-, and physical-level reconfigurations are needed. This UROCA architecture meets the requirements of portability, scalability, changeability and responsiveness of future reconfigurable manufacturing systems.

The GPF model does not include translational joints and different than six DOF kinematic structure. This problem motivated authors to develop the n-DOF Global

Kinematic Model (n-GKM). The development of the n-GKM opens a new area in the reconfigurable robotics theory [10]. For the symbolic calculation of the n-GKM dynamic equations, the recursive Newton-Euler algorithm is employed using the symbolic algebra package MAPLE 12 and Automatic Separation Method (ASM) [6, 11]. The dynamic model is named the Global Dynamic Model (n-GDM).

LITERATURE REVIEW

Today's manufacturing environment is dominated by market change and global competition. Manufacturing success and survival are characterized by low cost and high quality to meet market and customer demands. Reconfigurable manufacturing systems are identified as a means of improving the production outcome by quickly adjusting its production capacity and functionality in response to sudden market changes and needs [1]. The goal here is to ensure that individual machinery systems, such as robots and CNC machines, are capable of reconfiguration for different applications. A variety of modular self-reconfigurable robots have been investigated. SUPERBOT a robotic system presented in [12] is a deployable robot and build of modules. The modules have to connect and disconnect each other to form a new chain loop needed to perform a new task. The modules (units) of the modular and reconfigurable robots are independent units with their own actuation, power, CPU, memory and physical hardware connectors. In [13], autonomous modules (CONRO) are introduced with the capability to be reconfigured in different shapes, such as snakes or hexapods. Many modular and self-reconfigurable robots are investigated in [14-18]. These robots have the ability to change their number of links, form closed chains and open chains, and form single or multiple branches. All the discussed modular self-reconfigurable robots need special physical features (docking) to connect and release the modules. Furthermore, different applications are executed with a central control to direct and shape the final required structure, which enables the robot to perform a specified task. Accordingly, D-H parameters have not been introduced to describe the kinematic and dynamic models of the resulting configuration. For space explorations purposes, a conceptual design of a reconfigurable robot has been developed [19]. The authors proposed a non-modular reconfigurable robot with changing D-H parameters, such as the sliding link length and rotary angle. A modular and reconfigurable robot for industrial purposes has been introduced [20]. The PROFACTOR GmbH has presented a modular and reconfigurable robot with power cube (Mechatronical Components) modules. These modules are designed to be identical and self-contained with actuation, memory and mechanical, electrical and embedded programming. In [21], the authors derived the kinematic and dynamic models of reconfigurable robots using D-H parameters for different sets of joints, links and gripper modules. Furthermore, a library of modules is formed from which any module can be called with its associated kinematic and dynamic models. In [22], a modular and reconfigurable robot design is introduced with modular joints and links. The proposed design introduces zero links, offset to increase the robot's dexterity and maximizes the reachability. This link design methodology is producing singularity. A Task-based configuration optimization based on a genetic algorithm was used to solve a pre-defined set of modules for specific kinematic configuration [23]. In [4], the authors present a unified reconfigurable kinematic model and solver for 6R industrial robots. The reconfigurable kinematic model includes six rotational joints, each with two different positive directions controlled by reconfigurable parameters K_i as a function of the robot's twist angles. This model does not include the translational type of joints, which limits the model to represent only the 6R industrial robots.

To improve the capability of the previously developed model, a novel Global n-DOF Model was developed. This model includes rotational and translational joints that has six possible z-axis (axis of rotations and/or joints translations). This global model can be used to represent any open kinematic configuration machinery.

THE N-DOF GLOBAL KINEMATIC MODEL

The previously developed n-DOF Global Kinematic Model (n-GKM) represents a n-DOF kinematic structure with either rotational and/or translational joints [10]. For the n-GKM the D-H parameters are presented in Table **1**. The link twist angle α_i has only five different values, which maintain perpendicularity between a joint's coordinate frames.

Table 1: D-H parameters for n-GKM model

i	d_i	θ_i	a_i	α_i
1	$R_1 d_{DH1} + T_1 d_1$	$R_1 \theta_1 + T_1 \theta_{DH1}$	a_1	$0°, \pm 180°, \pm 90°$
2	$R_2 d_{DH2} + T_2 d_2$	$R_2 \theta_2 + T_2 \theta_{DH2}$	a_2	$0°, \pm 180°, \pm 90°$
...	...	...	...	...
n	$R_n d_{DHn} + T_n d_n$	$R_n \theta_n + T_n \theta_{DHn}$	a_n	$0°, \pm 180°, \pm 90°$

A devotement of the n-DOF Global Kinematic Model (n-GKM) is much needed for supporting any kinematic configuration presented in Fig. **1**, and the possible redundant kinematic structures that are intended to support more than 6-DOF.

All D-H parameters presented in Table **1** are not fixed values. They represent and satisfy properties of all possible machinery kinematic structures. This means that each joint has six different positive directions of rotation and/or translation. Any joint's vector z_{i-1} can be placed in positive and/or negative direction of x, y, and z axis of the Cartesian coordinate frame. This is expressed in equations (1) & (2).

Rotational Joints: $R_i = 1 \; and \; T_i = 0$ **(1)**

Translational Joints: $R_i = 0 \; and \; T_i = 1$ **(2)**

R_i and T_i are used to control the selection of joint type (rotational and/or translational). The orthogonality between the joint's coordinate frames is defined with twist angles α_i. Their sinus and cosines are defined as the joint's reconfigurable parameters (K_{Si} & K_{Ci}) and expressed in equations (3) & (4).

$K_{Si} = \sin \alpha_i$ **(3)**

$K_{Ci} = \cos \alpha_i$ **(4)**

To build a reconfigurable joint, we need to include all six different positive directions of rotation and/or translation. The procedure will start from the first coordinate frame by defining the orientation of the vector, Z_0. Because we have six combinations of vector-Z_0, we will start from the first one named, Z_0^1, see Fig.

1. The selection of vector Z_0^1, can be combined with four more orientations of vectors X_0 and Y_0. They are: X_0^{11}, Y_0^{11}, X_0^{12}, Y_0^{12}, X_0^{13}, Y_0^{13}, X_0^{14}, Y_0^{14}. The other combinations are generated similarly.

This will produce a reconfigurable joint to have 24 different possible coordinate frames. When we add to two possible types of joints, the total number of different combinations will be 48. The graphical representation of a single reconfigurable joint is given in Fig. **1**.

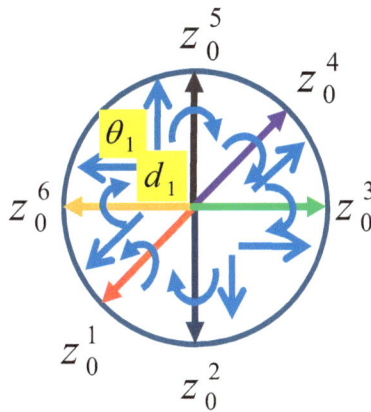

Figure 1: Reconfigurable model of Joint 1.

The total number of all possible kinematic structures, C_n is expressed in equation (5). Each combination can support either rotational or translational joint types.

$$C_n = 48(48-1)^{n-1} \tag{5}$$

The kinematics of the n-GKM can be calculated using the multiplication of the all homogeneous matrices from the base frame to the flange frame. The homogeneous transformation matrix for the n-DOF Global Kinematic Model (n-GKM) is shown in equation (6).

$$
{}^{i-1}A_i = \begin{bmatrix}
\cos(R_i\theta_i + T_i\theta_{DHi}) & -K_{Ci}\sin(R_i\theta_i + T_i\theta_{DHi}) & K_{Si}\sin(R_i\theta_i + T_i\theta_{DHi}) & a_i\cos(R_i\theta_i + T_i\theta_{DHi}) \\
\sin(R_i\theta_i + T_i\theta_{DHi}) & K_{Ci}\cos(R_i\theta_i + T_i\theta_{DHi}) & -K_{Si}\cos(R_i\theta_i + T_i\theta_{DHi}) & a_i\sin(R_i\theta_i + T_i\theta_{DHi}) \\
0 & K_{Si} & K_{Ci} & R_id_{DHi} + T_id_i \\
0 & 0 & 0 & 1
\end{bmatrix}
$$

$$i = 1, 2, \ldots, n \tag{6}$$

The kinematics and dynamic models for the n-GKM were developed previously [10] and [24]. Using the existing models, the global 2-DOF model will be used to illustrate the unified methodology and its applications for control processes.

THE 2-DOF GKM MODEL EXAMPLE

The D-H parameters of the 2-GKM kinematic model are presented in Table **2**.

Table 2: D-H parameters for 2-GKM model

i	d_i	θ_i	a_i	α_i
1	$R_1 d_{DH1} + T_1 d_1$	$R_1 \theta_1 + T_1 \theta_{DH1}$	a_1	$0°, \pm180°, \pm90°$
2	$R_2 d_{DH2} + T_2 d_2$	$R_2 \theta_2 + T_2 \theta_{DH2}$	a_2	$0°, \pm180°, \pm90°$

The reconfigurable parameters of the 2-GKM kinematic model are presented in Table **3**.

Table 3: D-H reconfigurable parameters for 2-GKM model

$K_{S1} = \sin(\pm90°) = \pm1$	$K_{C1} = \cos(\pm180°; 0) = \pm1$
$K_{S2} = \sin(\pm90°) = \pm1$	$K_{C2} = \cos(\pm180°; 0) = \pm1$

The 2-GKM kinematic model is graphically presented in Fig. **2**. Each joint is reconfigurable and in between any two joints all links have changeable values. The number of different kinematic structures can be calculated using equation (7), where n=2.

$$C_2 = 48\,(48-1)^{2-1} = 2256 \tag{7}$$

Forward Kinematics of 2-GKM Model

The two homogeneous transformation matrices $^0A_1$ and $^1A_2$ for the 2-GKM model are generated from the general homogeneous matrix (6). By multiplying these two matrices the arm transformation matrix can be calculated. See equation (8):

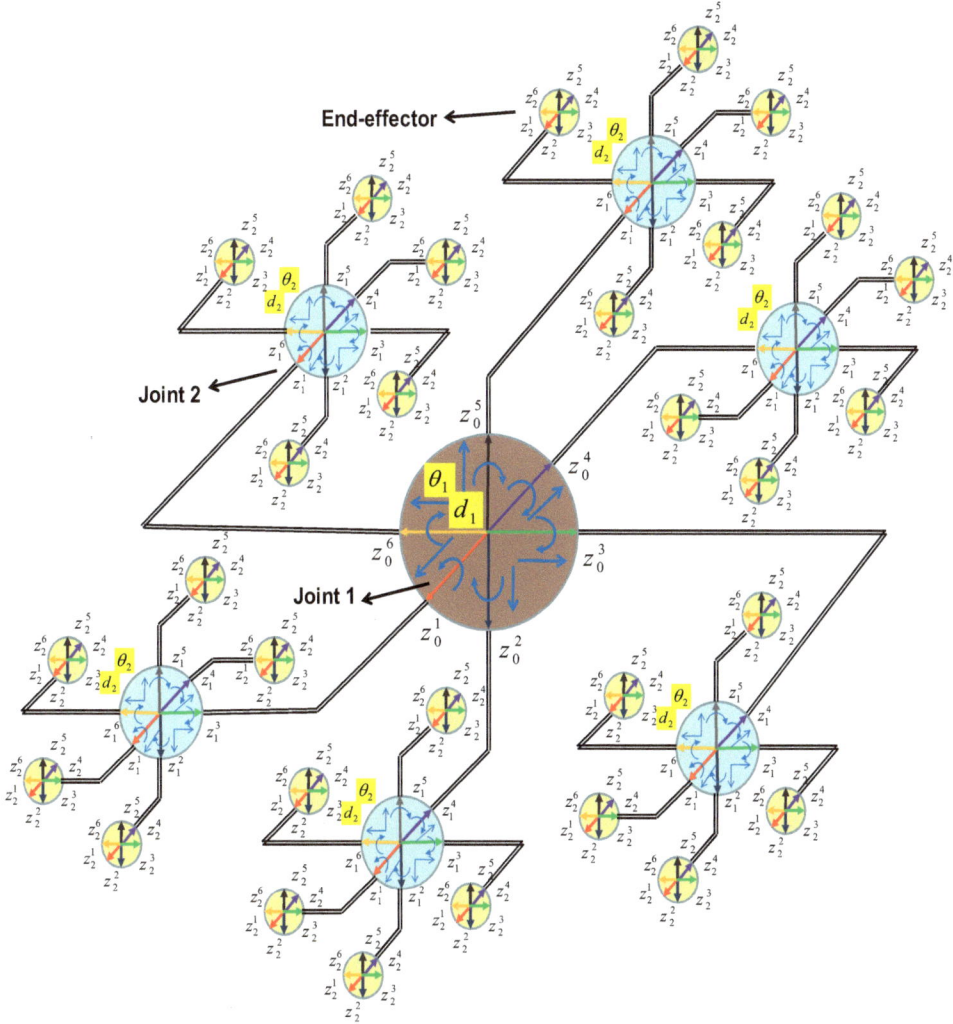

Figure 2: Kinematic Structure of the 2-GKM Model.

$$^{0}A_{1}\,^{1}A_{2} = \,^{0}A_{2} = \begin{bmatrix} n_x & s_x & a_x & p_x \\ n_y & s_y & a_y & p_y \\ n_z & s_z & a_z & p_z \\ 0 & 0 & 0 & 1 \end{bmatrix}$$

(8)

Equation (8) represents a forward kinematic solution for the 2-GKM model. The first 3×3 matrix is a rotational matrix, which represents the projections of the normal, sliding and approach vectors. The position of the end-effecter is expressed with 3×1 matrix. The elements of the transformation matrix are a function of the 2-GKM model's D-H parameters and they are presented in equations (9-20).

$$n_x = \cos[(R_1\theta_1 + T_1\theta_{DH1}) + K_{c1}(R_2\theta_2 + T_2\theta_{DH2})] \tag{9}$$

$$n_y = \sin[(R_1\theta_1 + T_1\theta_{DH1}) + K_{c1}(R_2\theta_2 + T_2\theta_{DH2})] \tag{10}$$

$$n_z = K_{s1}\sin(R_2\theta_2 + T_2\theta_{DH2}) \tag{11}$$

$$s_x = K_{c1}K_{c2}\sin[(R_1\theta_1 + T_1\theta_{DH1}) + K_{c1}(R_2\theta_2 + T_2\theta_{DH2})] + K_{s1}K_{s2}\sin(R_1\theta_1 + T_1\theta_{DH1}) \tag{12}$$

$$s_y = K_{c1}K_{c2}\cos[(R_1\theta_1 + T_1\theta_{DH1}) + K_{c1}(R_2\theta_2 + T_2\theta_{DH2})] - K_{s1}K_{s2}\cos(R_1\theta_1 + T_1\theta_{DH1}) \tag{13}$$

$$s_z = K_{s1}K_{c2}\cos(R_2\theta_2 + T_2\theta_{DH2}) + K_{c1}K_{s2} \tag{14}$$

$$a_x = K_{c1}K_{s2}\sin[(R_1\theta_1 + T_1\theta_{DH1}) + K_{c1}(R_2\theta_2 + T_2\theta_{DH2})] + K_{s1}K_{c2}\sin(R_1\theta_1 + T_1\theta_{DH1}) \tag{15}$$

$$a_y = -K_{c1}K_{s2}\cos[(R_1\theta_1 + T_1\theta_{DH1}) + K_{c1}(R_2\theta_2 + T_2\theta_{DH2})] - K_{s1}K_{c2}\cos(R_1\theta_1 + T_1\theta_{DH1}) \tag{16}$$

$$a_z = -K_{s1}K_{s2}\cos(R_2\theta_2 + T_2\theta_{DH2}) + K_{c1}K_{c2} \tag{17}$$

$$p_x = a_2\cos[(R_1\theta_1 + T_1\theta_{DH1}) + K_{c1}(R_2\theta_2 + T_2\theta_{DH2})] + \\ K_{s1}\sin(R_1\theta_1 + T_1\theta_{DH1})(R_2d_{DH2} + T_2d_2) + a_1\cos(R_1\theta_1 + T_1\theta_{DH1}) \tag{18}$$

$$p_y = a_2\sin[(R_1\theta_1 + T_1\theta_{DH1}) + K_{c1}(R_2\theta_2 + T_2\theta_{DH2})] - \\ K_{s1}\cos(R_1\theta_1 + T_1\theta_{DH1})(R_2d_{DH2} + T_2d_2) + a_1\sin(R_1\theta_1 + T_1\theta_{DH1}) \tag{19}$$

$$p_z = K_{s1}a_2\sin(R_2\theta_2 + T_2\theta_{DH2}) + K_{c1}(R_2d_{DH2} + T_2d_2) + (R_1d_{DH1} + T_1d_1) \tag{20}$$

Inverse Kinematics of 2-GKM Model

Combining equations (9)-(20) the inverse kinematic is calculated. First we calculated joint angles θ_1 and θ_2, and their solutions are expressed in equations (27) and (28).

$$(R_1\theta_1 + T_1\theta_{DH1}) + K_{c1}(R_2\theta_2 + T_2\theta_{DH2}) = A\tan 2(n_y, n_x) \tag{21}$$

$$p_x = a_2 n_x + K_{s1} \sin(R_1\theta_1 + T_1\theta_{DH1})(R_2 d_{DH2} + T_2 d_2) + a_1 \cos(R_1\theta_1 + T_1\theta_{DH1}) \tag{22}$$

$$p_y = a_2 n_y - K_{s1} \cos(R_1\theta_1 + T_1\theta_{DH1})(R_2 d_{DH2} + T_2 d_2) + a_1 \sin(R_1\theta_1 + T_1\theta_{DH1}) \tag{23}$$

$$\cos(R_1\theta_1 + T_1\theta_{DH1}) = \frac{p_x a_1 - a_1 a_2 n_x (R_2 d_{DH2} + T_2 d_2) K_{s1}(p_y - a_2 n_y)}{a_1^2 + K_{s1}^2 (R_2 d_{DH2} + T_2 d_2)^2} \tag{24}$$

$$\sin(R_1\theta_1 + T_1\theta_{DH1}) = \frac{p_y a_1 - a_1 a_2 n_y (R_2 d_{DH2} + T_2 d_2) K_{s1}(p_x - a_2 n_x)}{a_1^2 + K_{s1}^2 (R_2 d_{DH2} + T_2 d_2)^2} \tag{25}$$

$$R_1\theta_1 + T_1\theta_{DH1} = A\tan 2\left(\sin(R_1\theta_1 + T_1\theta_{DH1}), \cos(R_1\theta_1 + T_1\theta_{DH1}) \right) \tag{26}$$

$$\theta_1 = \frac{1}{R_1}\left(A\tan 2\left(\sin(R_1\theta_1 + T_1\theta_{DH1}), \cos(R_1\theta_1 + T_1\theta_{DH1}) \right) - T_1\theta_{DH1} \right) \tag{27}$$

$$\theta_2 = \frac{1}{R_2}\left\{ \frac{1}{K_{c1}}\left[A\tan 2(n_y, n_x) - (R_1\theta_1 + T_1\theta_{DH1}) \right] - T_2\theta_{DH2} \right\} * config \tag{28}$$

The 2-GKM model has two solutions of the inverse kinematic for rotational joints. These two configurations are UP and DOWN and they are graphically presented in Fig. **3**. $OJ_A P$ is a DOWN configuration and $OJ_B P$ an UP configuration. The mathematical calculation for the two configurations is presented in the equations (29-32). The projections of the center C is shown in Fig. **4**, and its coordinates are calculated using the equations (33-38).

$$C_y > 0 \quad and \quad C_y > J_y \quad \Rightarrow \quad config = DOWN \tag{29}$$

$$C_y > 0 \quad and \quad C_y < J_y \quad \Rightarrow \quad config = UP \tag{30}$$

$$C_y < 0 \quad and \quad C_y > J_y \quad \Rightarrow \quad config = DOWN \tag{31}$$

$$C_y < 0 \quad and \quad C_y < J_y \quad \Rightarrow \quad config = UP \tag{32}$$

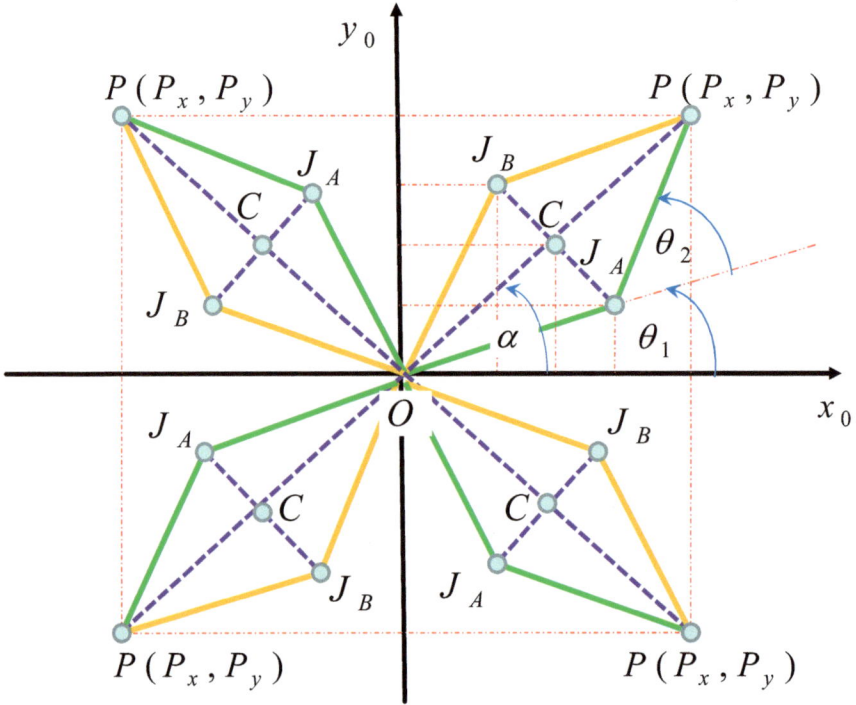

Figure 3: Different Configurations for the 2-GKM Model.

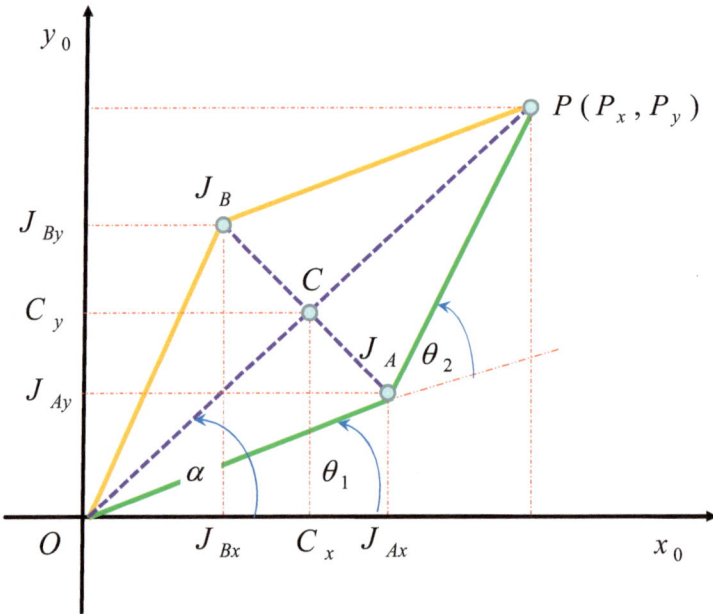

Figure 4: Projections of the center C.

$$C_x = OC \cos \alpha \tag{33}$$

$$\alpha = tg^{-1} \left(\frac{P_y}{P_x} \right) \tag{34}$$

$$OC = \frac{1}{2} OP = \frac{1}{2} \sqrt{P_x^2 + P_y^2} \tag{35}$$

$$C_x = \frac{1}{2} \sqrt{P_x^2 + P_y^2} \cos \left[tg^{-1} \left(\frac{P_y}{P_x} \right) \right] \tag{36}$$

$$C_y = OC \sin \alpha \tag{37}$$

$$C_y = \frac{1}{2} \sqrt{P_x^2 + P_y^2} \sin \left[tg^{-1} \left(\frac{P_y}{P_x} \right) \right] \tag{38}$$

Combining equation (19) and (20) the translational joints are calculated. Their results are expressed in equations (39) and (40).

$$d_2 = \frac{1}{T_2} \left(\frac{a_2 \sin[(R_1\theta_1 + T_1\theta_{DH1}) + K_{c1}(R_2\theta_2 + T_2\theta_{DH2})] + a_1 \sin(R_1\theta_1 + T_1\theta_{DH1}) - p_y}{K_{s1} \cos(R_1\theta_1 + T_1\theta_{DH1})} - R_2 d_{DH2} \right) \tag{39}$$

$$d_1 = T_1 \left\{ p_z - K_{s1} a_2 \sin(R_2\theta_2 + T_2\theta_{DH2}) - R_2 d_{DH1} - \frac{K_{c1} a_2 \sin[(R_1\theta_1 + T_1\theta_{DH1}) + K_{c1}(R_2\theta_2 + T_2\theta_{DH2})] K_{c1} a_1 \sin(R_1\theta_1 + T_1\theta_{DH1}) - p_y}{K_{s1}^2 \cos(R_1\theta_1 + T_1\theta_{DH1})} \right\} \tag{40}$$

The inverse kinematics solution for the 2GKM includes translational and rotational joint solutions. This must be expressed as generalized coordinates and their mathematical expression is given in the following equation:

$$q_i = R_i\theta_i + T_i d_i \tag{41}$$

Dynamics of 2-GKM Model

For the previously developed 2-GKM, the dynamic equations are solved. This dynamic model, called the 2-GDM (Global Dynamics Model), can be reconfigured from one dynamic group to another or from one application to another by using the configuration parameters as defined Table **3**. This model is graphically shown in Fig. **5**.

Each joint is presented with different coordinate systems. Each link has mass m_1 and m_2, and a center of mass P_{C1} and P_{C2}, respectively. From the Fig. **6**, we can see that P_{C1} is between Joint 1 and Joint 2. The coordinates of the P_{C1} are defined relative to the Joint 2 frame: x_1, y_1, z_1. In Fig. **6**, only one case of the four possible center of mass, P_{C1} is shown.

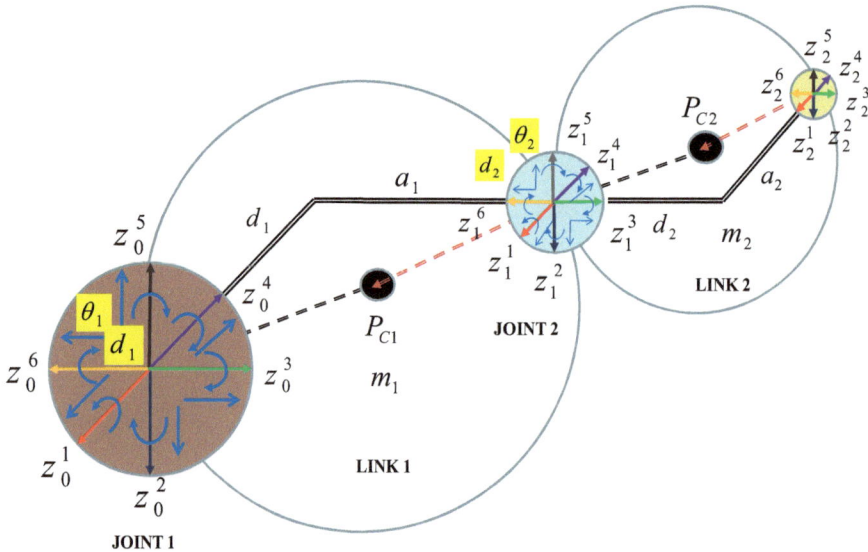

Figure 5: 2-DGM Model.

To find the centers of mass of each link for the 2-DGM model, we need to include all possible coordinate frames for both joints.

Starting from the Link 1 center of mass P_{C1}, development of the general coordinates has been determined by observing four different cases. This includes four possible centers of mass P_{C1}: P_{C1}^1, P_{C1}^2, P_{C1}^3, and P_{C1}^4. Each case is related to different coordinate frames of Joint 2. From Fig. **6** we can see that one selection of z_0^1 axis, can support four different x-axis: x_0^{11}, x_0^{12}, x_0^{13}, and x_0^{14}. For Joint 2, there are sixteen different combinations of x_1: x_1^1, x_1^2, x_1^3, x_1^4. Each of the four x-combinations has four more combinations of the Joint 2 coordinate frame. By observing the coordinates of each center of mass, P_{C1}^1, P_{C1}^2, P_{C1}^3, and P_{C1}^4, relative to all sixteen possible combinations, the general unified solution has been developed and presented in equation (42).

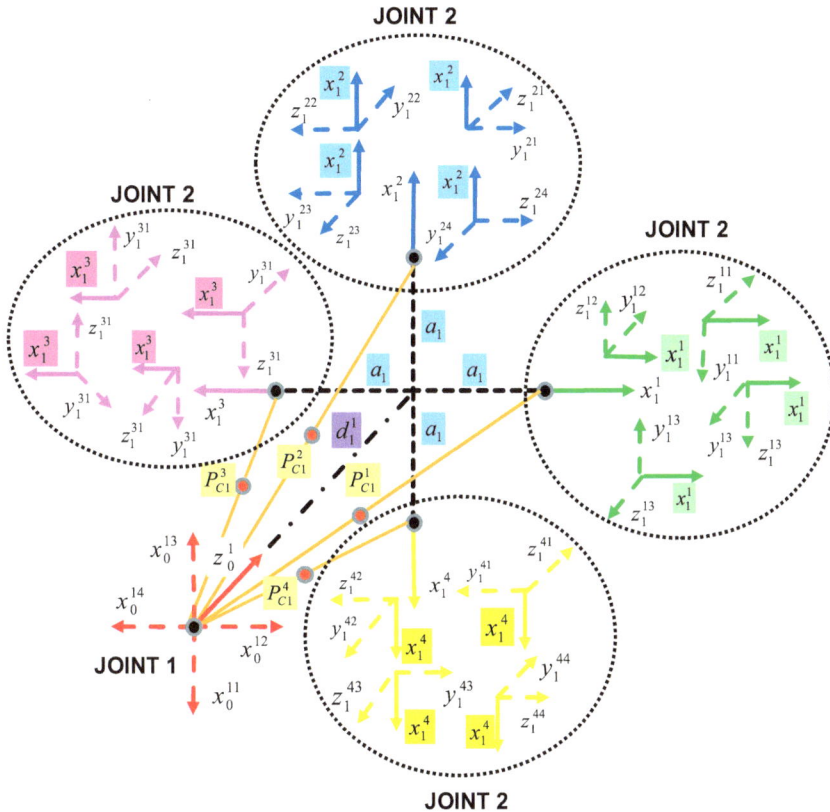

Figure 6: Center of mass for Link 1 of the 2-GDM Model.

$$P_{ci} = \begin{bmatrix} -\cos(\theta_{DHi})\dfrac{a_i}{2} \\[2mm] K_{ci}\sin(\theta_{DHi})\dfrac{a_i}{2} - K_{si}\dfrac{d_{DHi}}{2} \\[2mm] -K\sin(\theta_{DHi})\dfrac{a_i}{2} - K_{ci}\dfrac{d_{DHi}}{2} \end{bmatrix}, \ i = 1,2 \qquad (42)$$

The moment of inertia of the center of mass for each link is shown in equation (43).

$$I_i = \begin{bmatrix} I_{xi} & 0 & 0 \\ 0 & I_{yi} & 0 \\ 0 & 0 & I_{zi} \end{bmatrix}, \; i = 1, 2 \tag{43}$$

The 2-GDM model is calculated from the presented information. The equations of motion for a 2-DOF manipulator are given by equation (44).

$$A(q)\ddot{q} + B(q)[\dot{q}\dot{q}] + C(q)[\dot{q}^2] + G(q) = \tau \tag{44}$$

Equation (44) represents the dynamics of the ideal rigid bodies (links) connected with the joints, where there is a generalized force/torque vector τ acting at the each joint. In the case where we have 2(RT) joints, the vector τ is a torque or/and force about/along each axis of rotation/translation, produced by an actuator that is usually a DC motor [25].

Matrix A is the 2×2 inertia matrix, B is the 2×1 Coriolis torque matrix, C is the 2×2 centrifugal torque matrix, and G is a 2×1 gravity torque vector. Vector q is the vector of generalized joint coordinates: $q = \begin{bmatrix} q_1 & q_2 \end{bmatrix}^T$, where q_1 and q_2 are defined in equation (41). Vectors $\dot{q}$, $\ddot{q}$ are the vectors of joint velocities and acceleration respectively.

Vector $[\dot{q}\dot{q}]$ is the vector of velocity products and it is given that $[\dot{q}\dot{q}] = [\dot{q}_1\dot{q}_2]$. Vector $[\dot{q}^2] = \begin{bmatrix} \dot{q}_1^2 & \dot{q}_2^2 \end{bmatrix}$ is the vector of squared velocity.

The calculation of the 2-DGM model was done using the RNE (Recursive Newton-Euler) algorithm.

Forward Computation of Velocity and Accelerations

The joint coordinate frames (Fig. **2**) are assigned using the D-H notation (Table **2**) and are expressed with their homogeneous transformation matrices $^{i-1}A_i$ ($i = 1, 2$), as shown in equation (6). The upper $3X3$ sub matrices of the each homogeneous transformation matrices represent the rotational matrices for each joint, and they are presented in equations (45) and (46):

$$
{}^0R_1 = \begin{bmatrix} \cos(R_1\theta_1 + T_1\theta_{DH1}) & -K_{C1}\sin(R_1\theta_1 + T_1\theta_{DH1}) & K_{S1}\sin(R_1\theta_1 + T_1\theta_{DH1}) \\ \sin(R_1\theta_1 + T_1\theta_{DH1}) & K_{C1}\cos(R_1\theta_1 + T_1\theta_{DH1}) & -K_{S1}\cos(R_1\theta_1 + T_1\theta_{DH1}) \\ 0 & K_{S1} & K_{C1} \end{bmatrix} \quad (45)
$$

$$
{}^1R_2 = \begin{bmatrix} \cos(R_2\theta_2 + T_2\theta_{DH2}) & -K_{C2}\sin(R_2\theta_2 + T_2\theta_{DH2}) & K_{S2}\sin(R_2\theta_2 + T_2\theta_{DH2}) \\ \sin(R_2\theta_2 + T_2\theta_{DH2}) & K_{C2}\cos(R_2\theta_2 + T_2\theta_{DH2}) & -K_{S2}\cos(R_2\theta_2 + T_2\theta_{DH2}) \\ 0 & K_{S2} & K_{C2} \end{bmatrix} \quad (46)
$$

The next step is to find the transpose of all rotational matrices: $({}^0R_1)^T$ and $({}^1R_2)^T$. See equations (47) and (48).

$$
({}^0R_1)^T = \begin{bmatrix} \cos(R_1\theta_1 + T_1\theta_{DH1}) & \sin(R_1\theta_1 + T_1\theta_{DH1}) & 0 \\ -K_{C1}\sin(R_1\theta_1 + T_1\theta_{DH1}) & K_{C1}\cos(R_1\theta_1 + T_1\theta_{DH1}) & K_{S1} \\ K_{S1}\sin(R_1\theta_1 + T_1\theta_{DH1}) & -K_{S1}\cos(R_1\theta_1 + T_1\theta_{DH1}) & K_{C1} \end{bmatrix} \quad (47)
$$

$$
({}^1R_2)^T = \begin{bmatrix} \cos(R_2\theta_2 + T_2\theta_{DH2}) & \sin(R_2\theta_2 + T_2\theta_{DH2}) & 0 \\ -K_{C2}\sin(R_2\theta_2 + T_2\theta_{DH2}) & K_{C2}\cos(R_2\theta_2 + T_2\theta_{DH2}) & K_{S2} \\ K_{S2}\sin(R_2\theta_2 + T_2\theta_{DH2}) & -K_{S2}\cos(R_2\theta_2 + T_2\theta_{DH2}) & K_{C2} \end{bmatrix} \quad (48)
$$

The upper right $3X1$ sub matrices for each homogeneous transformation matrices represents the position vectors for each joint and is stated in equation (49) and (50):

$$
{}^0P_1 = \begin{bmatrix} a_1\cos(R_1\theta_1 + T_1\theta_{DH1}) \\ a_1\sin(R_1\theta_1 + T_1\theta_{DH1}) \\ R_1 d_{DH1} + T_1 d_1 \end{bmatrix} \quad (49)
$$

$$
{}^1P_2 = \begin{bmatrix} a_2\cos(R_2\theta_2 + T_2\theta_{DH2}) \\ a_2\sin(R_1\theta_1 + T_2\theta_{DH2}) \\ R_2 d_{DH2} + T_2 d_2 \end{bmatrix} \quad (50)
$$

The linear and angular velocity vectors and acceleration vectors for both joints are presented in equation (51).

$$
{}^{0}\dot{P}_{1} = \begin{bmatrix} 0 \\ 0 \\ \dot{d}_{1} \end{bmatrix}, \; {}^{1}\dot{P}_{2} = \begin{bmatrix} 0 \\ 0 \\ \dot{d}_{2} \end{bmatrix}, \; {}^{0}\dot{\theta}_{1} = \begin{bmatrix} 0 \\ 0 \\ \dot{\theta}_{1} \end{bmatrix}, \; {}^{1}\dot{\theta}_{2} = \begin{bmatrix} 0 \\ 0 \\ \dot{\theta}_{2} \end{bmatrix}, \; {}^{0}\ddot{P}_{1} = \begin{bmatrix} 0 \\ 0 \\ \ddot{d}_{1} \end{bmatrix}, \; {}^{1}\ddot{P}_{2} = \begin{bmatrix} 0 \\ 0 \\ \ddot{d}_{2} \end{bmatrix}, \; {}^{0}\ddot{\theta}_{1} = \begin{bmatrix} 0 \\ 0 \\ \ddot{\theta}_{1} \end{bmatrix},
$$

$$
{}^{1}\ddot{\theta}_{2} = \begin{bmatrix} 0 \\ 0 \\ \ddot{\theta}_{2} \end{bmatrix}
\tag{51}
$$

Using the appropriate rotation matrices and angular velocity vectors, the angular and linear velocities for rotational and translational joints can be calculated from equations (52) and (53) respectively:

$$
{}^{i}({}^{0}\omega_{i}) = {}^{i}R_{i-1}[{}^{i-1}({}^{0}\omega_{i-1}) + R_{i}({}^{i-1}\dot{\theta}_{i})], \; i = 1,2
\tag{52}
$$

$$
{}^{i}({}^{0}V_{i}) = {}^{i}R_{i-1}{}^{i-1}({}^{0}V_{i-1}) + R_{i}\left[{}^{i}({}^{0}\omega_{i}) \times {}^{i}R_{i-1}{}^{i-1}P_{i}\right] + R_{i}\left\{{}^{i}R_{i-1}\left[{}^{i-1}\dot{P}_{i} + {}^{i-1}({}^{0}\omega_{i-1}) \times {}^{i-1}P_{i}\right]\right\},
$$
$$
i = 1,2
\tag{53}
$$

The linear and angular acceleration for two joints is calculated from equations (54) and (55) respectively:

$$
{}^{i}({}^{0}\alpha_{i}) = {}^{i}R_{i-1}\left\{{}^{i-1}({}^{0}\alpha_{i-1}) + R_{i}\left[{}^{i-1}({}^{0}\omega_{i-1}) \times {}^{i-1}\dot{\theta}_{i} + {}^{i-1}\ddot{\theta}_{i}\right]\right\}, \; i = 1,2
\tag{54}
$$

$$
{}^{i}({}^{0}a_{i}) = {}^{i}R_{i-1}\{{}^{i-1}({}^{0}a_{i-1}) + {}^{i-1}({}^{0}\alpha_{i-1}) \times {}^{i-1}P_{i} + {}^{i-1}({}^{0}\omega_{i-1}) \times ({}^{i-1}({}^{0}\omega_{i-1}) \times {}^{i-1}P_{i})
$$
$$
+ R_{i}[2^{i-1}({}^{0}\omega_{i-1}) \times ({}^{i-1}\dot{\theta}_{i} \times {}^{i-1}P_{i}) + {}^{i-1}\ddot{\theta}_{i} \times {}^{i-1}P_{i} + {}^{i-1}\omega_{i} \times ({}^{i-1}\dot{\theta}_{i} \times {}^{i-1}P_{i})] + T_{i}[2^{i-1}({}^{0}\omega_{i-1}) \times {}^{i-1}\dot{P}_{i} + {}^{i-1}\ddot{P}_{i}]\}
$$
$$
i = 1,2
\tag{55}
$$

The linear acceleration of the center of mass is calculated from equation (56):

$$
{}^{i}({}^{0}a_{ci}) = {}^{i}({}^{0}a_{i}) + {}^{i}({}^{0}\alpha_{i}) \times {}^{i}({}^{i}P_{ci}) + {}^{i}({}^{0}\omega_{i}) \times [{}^{i}({}^{0}\omega_{i}) \times {}^{i}({}^{i}P_{ci})], \; i = 1,2
\tag{56}
$$

The radial distances to the centers of mass, P_{C1} and P_{C2}, are defined in equation (42).

Backward Computation of Velocity and Accelerations

The general gravity vectors for the 2-GDM model are expressed:

$$^{i}g = {}^{i}R_{0}{}^{0}g, \ i = 1,2 \tag{57}$$

From Fig. **5**, the gravity vector must be expressed in x, y and z directions:

$$^{0}g = \begin{bmatrix} G_{x}g \\ G_{y}g \\ G_{z}g \end{bmatrix}, \text{ where } g = 9.81 m/s^{2}. \text{ The parameters } G_{x}, \ G_{y}, \text{ and } G_{z} \text{ are used to}$$

control the gravity vector and they depend on the selected kinematic configuration.

Once the velocities and accelerations of the links are found, the joint forces and moments can be computed one link at a time starting from the end-effector link and ending at the base link. It is assumed that there is no load at the end-effecter; therefore, $^{2}({}^{2}f_{Tool}) = 0$, and $^{2}({}^{2}n_{Tool}) = 0$.

$$^{i}(f_{i}^{*}) = -m_{i}{}^{i}({}^{0}a_{ci}), \ i = 1,2 \tag{58}$$

$$^{i}(n_{i}^{*}) = -{}^{i}I_{i}{}^{i}({}^{0}\alpha_{ci}) - {}^{i}({}^{0}\omega_{ci}) \times [{}^{i}I_{i}{}^{i}({}^{0}\omega_{ci})], \ i = 1,2 \tag{59}$$

$^{i}I_{i}$ is the inertia tensor. This is expressed simply in equation (43).

The force and moment balance equations about the center of mass of link i in recursive form can be written as:

$$^{i}(f_{i}^{*}) + {}^{i}({}^{i-1}f_{i}) - {}^{i}({}^{i}f_{i+1}) + m_{i}{}^{i}g = 0, \ i = 1,2 \tag{60}$$

$$^{i}(n_{i}^{*}) + {}^{i}({}^{i-1}n_{i}) - {}^{i}({}^{i}n_{i+1}) - [{}^{i}({}^{i-1}P_{i}) + {}^{i}({}^{i}P_{ci})] \times {}^{i}({}^{i-1}f_{i}) + {}^{i}({}^{i}P_{ci}) \times {}^{i}({}^{i}f_{i+1}) = 0, i = 1,2 \tag{61}$$

Once the reaction forces and moments are computed in the ith link frame, they are converted into the $(i-1)^{th}$ link frame by the following equations:

$$^{i}({}^{i-1}f_{i}) = {}^{i-1}R_{i}{}^{i}({}^{i-1}f_{i}), \ i = 1,2 \tag{62}$$

$$^{i}({}^{i-1}n_{i}) = {}^{i-1}R_{i}{}^{i}({}^{i-1}n_{i}), \ i = 1,2 \tag{63}$$

Actuator torques and forces τ_{i} are obtained by projecting the forces onto their corresponding joint axes:

$$\tau_i = {}^i\left({}^{i-1}n_i\right)^{T}{}^{i-1}z_{i-1}, \ i=1,2 \tag{64}$$

The RNE procedure produced the final six expressions of the actuators' torque and force τ_i, $i=1,2$. Each of these six equations contain sums of products of matrices elements A, B, C, G, and trigonometric terms. To be able to get a dynamic equation in the form of equation (44), you need to generate each matrix A, B, C, and G, which means you need to calculate their elements: a_{11}, a_{12}, a_{21}, a_{22}, b_{112}, c_{11}, c_{12}, c_{21}, c_{22}, g_1 and g_2. To avoid complications when factoring out each element in each τ_i expression, the Automatic Separation Procedure (ASP) is used, which produces an automatic generation of the matrix elements. This method is explained in the next section.

Automatic Separation Method (ASM)

To avoid calculation complexity and reach the final goal of the dynamic equation in the form of equation (65), the Automatic Separation Procedure (ASP) was used. This procedure has three steps.

The first step is to simplify and organize the angular and linear velocity equations. To overcome these problems it is necessary to implement the basic trigonometric rules expressed in equations (65 – 67):

$$\sin\theta_i \cos\theta_j + K_i \cos\theta_i \sin\theta_j = \sin(\theta_i + K_i\theta_j), \ i=1,2, \ j=1,2, \ K_i = \pm 1 \tag{65}$$

$$\cos\theta_i \cos\theta_j - K_i \sin\theta_i \sin\theta_j = \cos(\theta_i + K_i\theta_j), \ i=1,2, \ j=1,2, \ K_i = \pm 1 \tag{66}$$

$$\sin^2\theta_i + \cos^2\theta_i = 1, \ i=1,2 \tag{67}$$

Two more simplifications can be applied, as shown in equations (68) and (69).

$$K_{si}{}^2 = 1 \ and \ K_{ci}{}^2 = 1, \ i=1,2 \tag{68}$$

$$K_{si}{}^3 = K_{si} \ and \ K_{ci}{}^3 = K_{ci}, \ i=1,2 \tag{69}$$

To be able to apply equations (68) and (69) using MAPLE 12®, the command "algsubs", must be used. This command substitutes sub-expressions into an expression. In equation (70) is an example of simplifying a_{11}, the first element of matrix A is given:

$$\text{algsubs}\,(K_6^3 = K_6, a_{11}) \tag{70}$$

The second step consists of ordering parameters of each element of the equation, which will help to continue the calculation without much complexity. The form is given in equation (71).

$$ELEMENT = K_i a_i d_i \sin\theta_i \cos\theta_i\,, i = 1,2 \tag{71}$$

The third step is the separation of elements and it is important for further calculations. This method is first applied to the angular acceleration $^i({}^0\alpha_i)_i$ and linear acceleration of the center of mass $^i({}^0a_{ci})$, such that these expressions are written as a sum of separated elements, as a function of the parameters of the vectors $[\ddot{q}]$, $[\dot{q}\dot{q}]$, $[\dot{q}^2]$. The angular acceleration $^i({}^0\alpha_i)_i$ is separated and expressed in equations (72-78).

$$^i({}^0\alpha_i)_i = {}^i({}^0\alpha_i)_{i(\alpha_i)} + {}^i({}^0\alpha_i)_{i(\ddot{d}_i)} + {}^i({}^0\alpha_i)_{i(\omega_i^2)} + {}^i({}^0\alpha_i)_{i(\dot{d}_i^2)} + {}^i({}^0\alpha_i)_{i(\omega_{ij})} + {}^i({}^0\alpha_i)_{i(\omega_i d_j)} \tag{72}$$

$$^i({}^0\alpha_i)_{i(\alpha_i)} = \begin{bmatrix} {}^i({}^0\alpha_i)_{i(\alpha_i)x} \\ {}^i({}^0\alpha_i)_{i(\alpha_i)y} \\ {}^i({}^0\alpha_i)_{i(\alpha_i)z} \end{bmatrix}, \; i = 1,2 \tag{73}$$

$$^i({}^0\alpha_i)_{i(\ddot{d}_i)} = \begin{bmatrix} {}^i({}^0\alpha_i)_{i(\ddot{d}_i)x} \\ {}^i({}^0\alpha_i)_{i(\ddot{d}_i)y} \\ {}^i({}^0\alpha_i)_{i(\ddot{d}_i)z} \end{bmatrix}, \; i = 1,2 \tag{74}$$

$$^i({}^0\alpha_i)_{i(\omega_i^2)} = \begin{bmatrix} {}^i({}^0\alpha_i)_{i(\omega_i^2)x} \\ {}^i({}^0\alpha_i)_{i(\omega_i^2)y} \\ {}^i({}^0\alpha_i)_{i(\omega_i^2)z} \end{bmatrix}, \; i = 1,2 \tag{75}$$

$$^{i}(^{0}\alpha_{i})_{i(d_{i}^{2})} = \begin{bmatrix} ^{i}(^{0}\alpha_{i})_{i(d_{i}^{2})x} \\ ^{i}(^{0}\alpha_{i})_{i(d_{i}^{2})y} \\ ^{i}(^{0}\alpha_{i})_{i(d_{i}^{2})z} \end{bmatrix}, \; i = 1,2 \tag{76}$$

$$^{i}(^{0}\alpha_{i})_{i(\omega_{ij})} = \begin{bmatrix} ^{i}(^{0}\alpha_{i})_{i(\omega_{ij})x} \\ ^{i}(^{0}\alpha_{i})_{i(\omega_{ij})y} \\ ^{i}(^{0}\alpha_{i})_{i(\omega_{ij})z} \end{bmatrix}, \; i = 1,2, \; j = 1,2 \tag{77}$$

$$^{i}(^{0}\alpha_{i})_{i(\omega_{i}d_{j})} = \begin{bmatrix} ^{i}(^{0}\alpha_{i})_{i(\omega_{i}d_{j})x} \\ ^{i}(^{0}\alpha_{i})_{i(\omega_{i}d_{j})y} \\ ^{i}(^{0}\alpha_{i})_{i(\omega_{i}d_{j})z} \end{bmatrix}, \; i = 1,2, \; j = 1,2 \tag{78}$$

The linear acceleration of the center of mass $^{i}(^{0}a_{ci})$ is separated and expressed in equations (79-85).

$$^{i}(^{0}a_{ci})_{i} = {}^{i}(^{0}a_{ci})_{i(\alpha_{i})} + {}^{i}(^{0}a_{ci})_{i(\ddot{d}_{i})} + {}^{i}(^{0}a_{ci})_{i(\omega_{i}^{2})} + {}^{i}(^{0}a_{ci})_{i(d_{i}^{2})} + {}^{i}(^{0}a_{ci})_{i(\omega_{ij})} + {}^{i}(^{0}a_{ci})_{i(\omega_{i}d_{j})} \tag{79}$$

$$^{i}(^{0}a_{ci})_{i(\alpha_{i})} = \begin{bmatrix} ^{i}(^{0}a_{ci})_{i(\alpha_{i})x} \\ ^{i}(^{0}a_{ci})_{i(\alpha_{i})y} \\ ^{i}(^{0}a_{ci})_{i(\alpha_{i})z} \end{bmatrix}, \; i = 1,2 \tag{80}$$

$$^{i}(^{0}a_{ci})_{i(\ddot{d}_{i})} = \begin{bmatrix} ^{i}(^{0}a_{ci})_{i(\ddot{d}_{i})x} \\ ^{i}(^{0}a_{ci})_{i(\ddot{d}_{i})y} \\ ^{i}(^{0}a_{ci})_{i(\ddot{d}_{i})z} \end{bmatrix}, \; i = 1,2 \tag{81}$$

$$^{i}(^{0}a_{ci})_{i(\omega_{i}^{2})} = \begin{bmatrix} ^{i}(^{0}a_{ci})_{i(\omega_{i}^{2})x} \\ ^{i}(^{0}a_{ci})_{i(\omega_{i}^{2})y} \\ ^{i}(^{0}a_{ci})_{i(\omega_{i}^{2})z} \end{bmatrix}, \; i = 1,2 \tag{82}$$

$$
{}^{i}({}^{0}a_{ci})_{i(d_{i}^{2})} = \begin{bmatrix} {}^{i}({}^{0}a_{ci})_{i(d_{i}^{2})x} \\ {}^{i}({}^{0}a_{ci})_{i(d_{i}^{2})y} \\ {}^{i}({}^{0}a_{ci})_{i(d_{i}^{2})z} \end{bmatrix}, \ i = 1, 2 \tag{83}
$$

$$
{}^{i}({}^{0}a_{ci})_{i(\omega_{ij})} = \begin{bmatrix} {}^{i}({}^{0}a_{ci})_{i(\omega_{ij})x} \\ {}^{i}({}^{0}a_{ci})_{i(\omega_{ij})y} \\ {}^{i}({}^{0}a_{ci})_{i(\omega_{ij})z} \end{bmatrix}, \ i = 1, 2, \ j = 1, 2 \tag{84}
$$

$$
{}^{i}({}^{0}a_{ci})_{i(\omega_{i}d_{j})} = \begin{bmatrix} {}^{i}({}^{0}a_{ci})_{i(\omega_{i}d_{j})x} \\ {}^{i}({}^{0}a_{ci})_{i(\omega_{i}d_{j})y} \\ {}^{i}({}^{0}a_{ci})_{i(\omega_{i}d_{j})z} \end{bmatrix}, \ i = 1, 2, \ j = 1, 2 \tag{85}
$$

The calculation of the inverse dynamics can be done such that all elements of the matrices from equation (44) can be automatically generated. Using equations (65-85), the automatic generation of all matrix elements can be produced, as shown below:

$$
a_{ij} = {}^{i}(R_{i}^{\ i-1}(n_{i(\alpha_{i})} + n_{i(\ddot{d}_{i})}) + T_{i}^{\ i-1}(f_{i(\alpha_{i})} + f_{i(\ddot{d}_{i})}))^{T\ i-1}z_{i-1}, \ i = 1, 2, \ j = 1, 2 \tag{86}
$$

$$
b_{i,ij} = {}^{i}(R_{i}^{\ i-1}(n_{i(\omega_{ij})} + n_{i(\omega_{i}\dot{d}_{j})}) + T_{i}^{\ i-1}(f_{i(\omega_{ij})} + f_{i(\omega_{i}\dot{d}_{j})}))^{T\ i-1}z_{i-1}, \ i = 1, 2, \ j = 1, 2 \tag{87}
$$

$$
c_{ij} = {}^{i}(R_{i}^{\ i-1}(n_{i(\omega_{i}^{2})} + n_{i(\dot{d}_{i}^{2})}) + T_{i}^{\ i-1}(f_{i(\omega_{i}^{2})} + f_{i(\dot{d}_{i}^{2})}))^{T\ i-1}z_{i-1}, \ i = 1, 2, \ j = 1, 2 \tag{88}
$$

$$
g_{i} = {}^{i}(R_{i}^{\ i-1}n_{i(GR_{i})} + T_{i}^{\ i-1}f_{i(GR_{i})})^{T\ i-1}z_{i-1}, \ i = 1, 2, \ j = 1, 2 \tag{89}
$$

From equations (86-89) all four matrices A, B, C, and G can be assembled. By selecting any robot that belongs to the 2-GKM model and entering all essential information, the dynamic model can be automatically generated. This provides a fundamental basis for subsequent control strategies.

$$\begin{bmatrix} a_{11} & a_{12} \\ a_{21} & a_{22} \end{bmatrix} \begin{bmatrix} \ddot{q}_1 \\ \ddot{q}_2 \end{bmatrix} + \begin{bmatrix} b_{112} \\ b_{212} \end{bmatrix} [\dot{q}_1 \dot{q}_2] + \begin{bmatrix} c_{11} & c_{12} \\ c_{21} & c_{22} \end{bmatrix} \begin{bmatrix} \dot{q}_1^2 \\ \dot{q}_2^2 \end{bmatrix} + \begin{bmatrix} G_1 \\ G_2 \end{bmatrix} = \begin{bmatrix} \tau_1 \\ \tau_2 \end{bmatrix} \tag{90}$$

The generalized coordinates q_i are expressed in equation (41). Each link is treated separately and their equations are as follows:

$$a_{11}\ddot{q}_1 + a_{12}\ddot{q}_2 + b_{112}\dot{q}_1\dot{q}_2 + c_{11}\dot{q}_1^2 + c_{12}\dot{q}_2^2 + G_1 = \tau_1 \tag{91}$$

$$a_{21}\ddot{q}_1 + a_{22}\ddot{q}_2 + b_{212}\dot{q}_1\dot{q}_2 + c_{21}\dot{q}_1^2 + c_{22}\dot{q}_2^2 + G_2 = \tau_2 \tag{92}$$

CONTROL PLATFORM FOR 2-GKM MODEL

The complexity of the full 2-GKM model affects the dynamic equations complexity and the block diagram. To be able to demonstrate how to make a control platform and to develop a controller, the sub model is selected from the 2-GKM. The selected model is 2-Ttr (2 DOF Rotational and rotational/translational joints).

The D-H parameters of the 2-Ttr kinematic model are presented in Table **3**.

Table 3: D-H parameters for 2-Ttr model

i	d_i	θ_i	a_i	α_i
1	d_1	$\theta_{DH1} = 90°$	0	90°
2	$R_2 d_{DH2} + T_2 d_2$	$R_2\theta_2 + T_2\theta_{DH2}$	0	0°

The reconfigurable parameters of the 2-Ttr kinematic model are presented in Table **4**.

Table 4: D-H reconfigurable parameters for 2-Ttr model

$K_{S1} = 1$	$K_{C1} = 0$	$R_1 = 0$	$T_1 = 1$
$K_{S2} = 0$	$K_{C2} = 1$	$R_2 = 1$	$T_2 = 1$

Using the information from Tables **3** and **4** and equations (45-92), the 2-Ttr mode is developed and its forward and inverse dynamic has been calculated using the ASM method.

The 2- Ttr kinematic model is graphically presented in Fig. **7**. The second joint is reconfigurable and in between any two joints all links have changeable values.

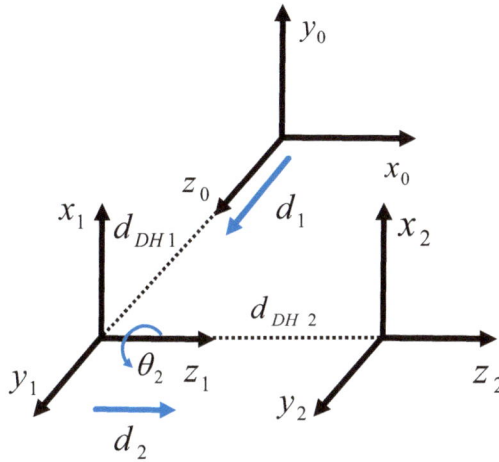

Figure 7: Kinematic Structure of the 2-Ttr Model.

The generalized joint coordinates for the 2-Ttr model are: $q = \begin{bmatrix} q_1 & q_2 \end{bmatrix}^T$, were $q_1 = d_1$ and $q_2 = R_2\theta_2 + T_2 d_2$.

Homogeneous transformation matrices are as follows:

$$ {}^0A_1 = \begin{bmatrix} 0 & 0 & 1 & 0 \\ 1 & 0 & 0 & 0 \\ 0 & 1 & 0 & d_1 \\ 0 & 0 & 0 & 1 \end{bmatrix} \tag{93} $$

$$ {}^1A_2 = \begin{bmatrix} \cos(R_2\theta_2) & -\sin(R_2\theta_2) & 0 & 0 \\ \sin(R_2\theta_2) & \cos(R_2\theta_2) & 0 & 0 \\ 0 & 0 & 1 & R_2 d_{DH2} + T_2 d_2 \\ 0 & 0 & 0 & 1 \end{bmatrix} \tag{94} $$

The 2-Ttr dynamic model is calculated using the Newton-Euler recursive calculation and ASM method. The force/torque equation is:

$$\begin{bmatrix} F_1 \\ F_2 T_2 + T_{L2} T_2 \end{bmatrix} = \begin{bmatrix} a_{11} & a_{12} \\ a_{21} & a_{22} \end{bmatrix} \begin{bmatrix} \ddot{d}_1 \\ R_2 \ddot{\theta}_2 + T_2 \ddot{d}_2 \end{bmatrix} + \begin{bmatrix} b_{112} \\ b_{212} \end{bmatrix} [\dot{d}_1 (R_2 \dot{\theta}_2 + T_2 \dot{d}_2)] + \begin{bmatrix} c_{11} & c_{12} \\ c_{21} & c_{22} \end{bmatrix} \begin{bmatrix} \dot{d}_1^2 \\ (R_2 \dot{\theta}_2 + T_2 \dot{d}_2)^2 \end{bmatrix} + \begin{bmatrix} G_1 \\ G_2 \end{bmatrix}$$

(95)

Each link is treated separately and their equations are as follows:

$$a_{11} \ddot{d}_1 + a_{12} \left(R_2 \ddot{\theta}_2 + T_2 \ddot{d}_2 \right) + b_{112} [\dot{d}_1 \left(R_2 \dot{\theta}_2 + T_2 \dot{d}_2 \right)] + c_{11} \dot{d}_1^2 + c_{12} \left(R_2 \dot{\theta}_2 + T_2 \dot{d}_2 \right)^2 + G_1 = \tau_1 \quad \textbf{(96)}$$

$$a_{21} \ddot{d}_1 + a_{22} \left(R_2 \ddot{\theta}_2 + T_2 \ddot{d}_2 \right) + b_{212} [\dot{d}_1 \left(R_2 \dot{\theta}_2 + T_2 \dot{d}_2 \right)] + c_{21} \dot{d}_1^2 + c_{22} \left(R_2 \dot{\theta}_2 + T_2 \dot{d}_2 \right)^2 + G_2 = \tau_2$$

(97)

The elements are given as follows:

$$a_{11} = \ddot{d}_1 (m_1 + m_2), \quad a_{12} = R_2 T_2 m_2 d_{DH2} \sin(R_2 \theta_2) \sin(\theta_2) \ddot{d}_2 \tag{98}$$

$$b_{112} = 0 \tag{99}$$

$$G_1 = m_2 \sin(R_2 \theta_2) g \tag{100}$$

$$c_{11} = 0, \quad c_{12} = \frac{1}{2} R_2^3 d_{DH2}^2 m_2 \sin(R_2 \theta_2) \sin(\theta_2) \omega_2^2 \tag{101}$$

$$a_{21} = 0, \quad a_{22} = m_2 T_2^2 \ddot{d}_2 \tag{102}$$

$$b_{212} = 0 \tag{103}$$

$$c_{21} = 0, \quad c_{22} = \frac{1}{2} T_2 R_2^2 m_2 d_{DH2} \omega_2^2 \tag{104}$$

$$G_2 = -R_2 d_{DH2} m_2 \sin(\theta_2) g \tag{105}$$

Considering DC motors in series with a gear train with (gear ratio n) and connected to a manipulator link, for either rotational or translational joints is presented in Figs. **8** and **9** respectively. The DC motors parameters are selected by Electro Craft E22 series and are given in Table **5**.

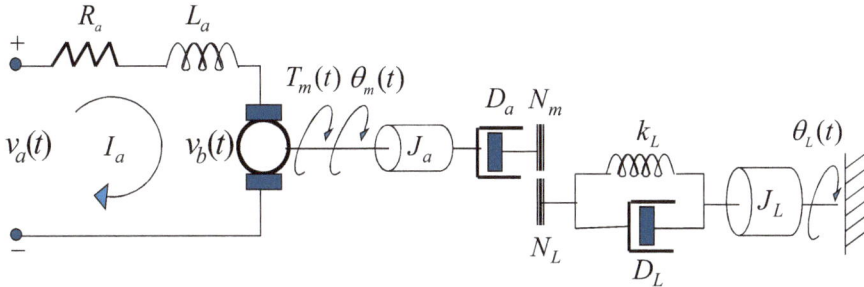

Figure 8: Lumped model of rotational link with DC Motor and gear.

The DC motor equations are as follows:

Motor torque: $T_m(s) = K_m I_a(s)$, $T_m(s) = T_L(s) + T_d(s)$, $T_d(s) = 0$ **(106)**

Armature voltage: $V_a(s) = R_a I_a(s) + sL_a I_a(s) + V_b(s)$ **(107)**

BMF voltage: $V_b(s) = K_b \omega(s)$, $\omega(s) = s\theta_m(s)$ **(108)**

Load torque: $T_L(s) = (J_L s^2 + D_L s + k_L)\theta_L(s)$ **(109)**

The transfer function is given in equation (110).

$$\frac{\theta_L(s)}{V_a(s)} = \frac{K_m \dfrac{N_m}{N_L} \dfrac{1}{J_m L_a}}{\left[s^3 + \dfrac{(J_m R_a + D_m L_a)}{J_m L_a} s^2 + \dfrac{(D_m R_a + K_b K_m + k_m L_a)}{J_m L_a} s + \dfrac{k_m}{J_m L_a} R_a \right]}$$ **(110)**

The DC motor equations are as follows:

Armature voltage: $V_a(s) = R_a I_a(s) + sL_a I_a(s) + V_b(s)$ **(111)**

BMF voltage: $V_b(s) = K_b \omega(s)$, $\omega(s) = s\theta_m(s)$ **(112)**

Gear ratio: $n = \dfrac{N_m}{N_L}$ (113)

Motor torque: $T_m(s) = K_m I_a(s)$, $T_m(s) = T_L(s)n$ (114)

Load torque: $T_L(s) = (J_L s^2 + D_L s)\theta_L(s) + rF(s)$ (115)

Translational Link: $F(s) = (ms^2 + D_T s)X(s)$ (116)

Mass displacement: $X(s) = r\theta_L(s)$ (117)

Load inertia and damping: $J_L = J_a(\dfrac{1}{n})^2 + J$, $D_L = D_a(\dfrac{1}{n})^2 + D$ (118)

The transfer function is given in equation (119).

$$\frac{X(s)}{V_a(s)} = \frac{\dfrac{nrK_m}{\left[n^2 L_a\left(J_L + r^2 m\right)\right]}}{s^3 + s^2 \dfrac{\left[n^2 R_a\left(J_L + r^2 m\right) + L_a\left(D_L + r^2 D_T\right)\right]}{\left[n^2 L_a\left(J_L + r^2 m\right)\right]} + s \dfrac{\left[n^2 R_a\left(D_L + r^2 D_T\right) + K_b K_m\right]}{\left[n^2 L_a\left(J_L + r^2 m\right)\right]}}$$

(119)

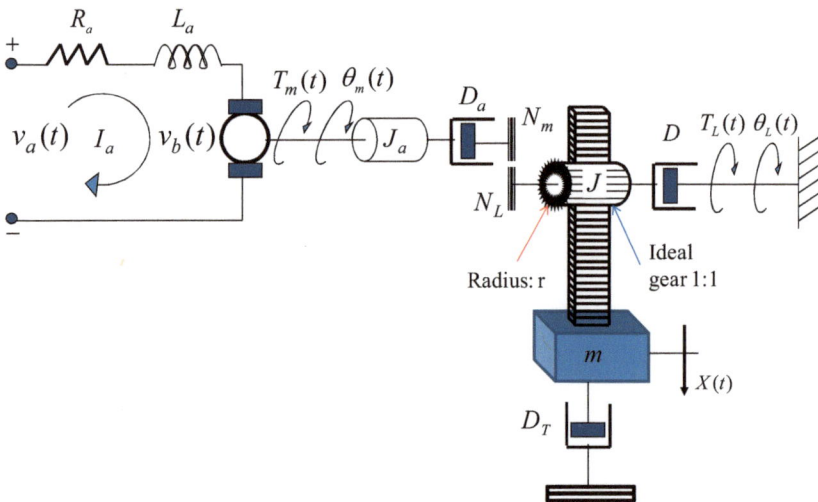

Figure 9: Lumped model of translational link with DC Motor and gear.

Table 5: DC Motors Information

	DC Motor 1	DC Motor 2
$V_a \, [kgm^2]$	24	24
$J_a \, [kgm^2]$	$98.9 * 10^{-7}$	$89.9 * 10^{-7}$
$D_a \, [\dfrac{Nms}{rad}]$	0.00148	0.000817
$R_a \, [ohm]$	0.4	0.4
$L_a \, [\dfrac{ohm}{s}]$	0.0007	0.0007
$I_a \, [A]$	$T_{stall1} \big/ RPM_1$	$T_{stall2} \big/ RPM_2$
$T_{stall} \, [Nm]$	0.1412	0.1412
$K_{m1} \, [\dfrac{Nm}{A}]$	$R_{a1} \big/ (V_{max} T_{stall1})$	$R_{a2} \big/ (V_{max} T_{stall2})$
$T_b \, [Nm]$	6.3	5.5
RPM	15000	15000
n	1/99.1233	1/98.3251

These two dynamic systems need to be coupled to present more realistic behaviour of different robotic systems, equation (120).

$$
\begin{bmatrix} F_1 \\ F_2 T_2 + T_{L2} R_2 \end{bmatrix} = \begin{bmatrix} \dfrac{J_{a1}}{r_1^2} & 0 \\ 0 & \dfrac{J_{a2}}{r_1^2} T_2 + J_{a2} R_2 \end{bmatrix} \begin{bmatrix} \ddot{\theta}_1 \\ R_2 \ddot{\theta}_2 + T_2 \ddot{d}_2 \end{bmatrix} + \begin{bmatrix} D_{a1} & 0 \\ 0 & D_{a2} \end{bmatrix} \begin{bmatrix} \dot{\theta}_1 \\ R_2 \dot{\theta}_2 + T_2 \dot{d}_2 \end{bmatrix} +
$$

$$
\begin{bmatrix} r_1 n_1^2 & r_2 n_2^2 T_2 + n_2^2 R_2 \end{bmatrix} \left\{ \begin{bmatrix} a_{11} & a_{12} \\ a_{21} & a_{22} \end{bmatrix} \begin{bmatrix} \ddot{\theta}_1 \\ R_2 \ddot{\theta}_2 + T_2 \ddot{d}_2 \end{bmatrix} + \begin{bmatrix} b_{112} \\ b_{212} \end{bmatrix} [\dot{\theta}_1 (R_2 \dot{\theta}_2 + T_2 \dot{d}_2)] + \begin{bmatrix} c_{11} & c_{12} \\ c_{21} & c_{22} \end{bmatrix} \begin{bmatrix} \dot{\theta}_1^2 \\ (R_2 \dot{\theta}_2 + T_2 \dot{d}_2)^2 \end{bmatrix} + \begin{bmatrix} G_1 \\ G_2 \end{bmatrix} \right\}
$$

$$\tag{120}$$

The importance of coupling the robot manipulator dynamics and robot joint actuators is presented in [26]. Using the equations for the robot links and motor's dynamics, and the motor information in Table **5**, the block diagram is generated in

Fig. **10**. The block diagram shows links dynamics (force/torque), two DC motors for each joint, and two controllers (PID).

For simple analysis, the step signal is used and the system response for each joint position (Linear/ Angular) is presented in Fig. **11**. It is clear that the PID controllers produce fast and accurate responses.

The simulation is done for the selected parameters: $R_2 = 0$ and $T_2 = 1$, which means that second link is translational.

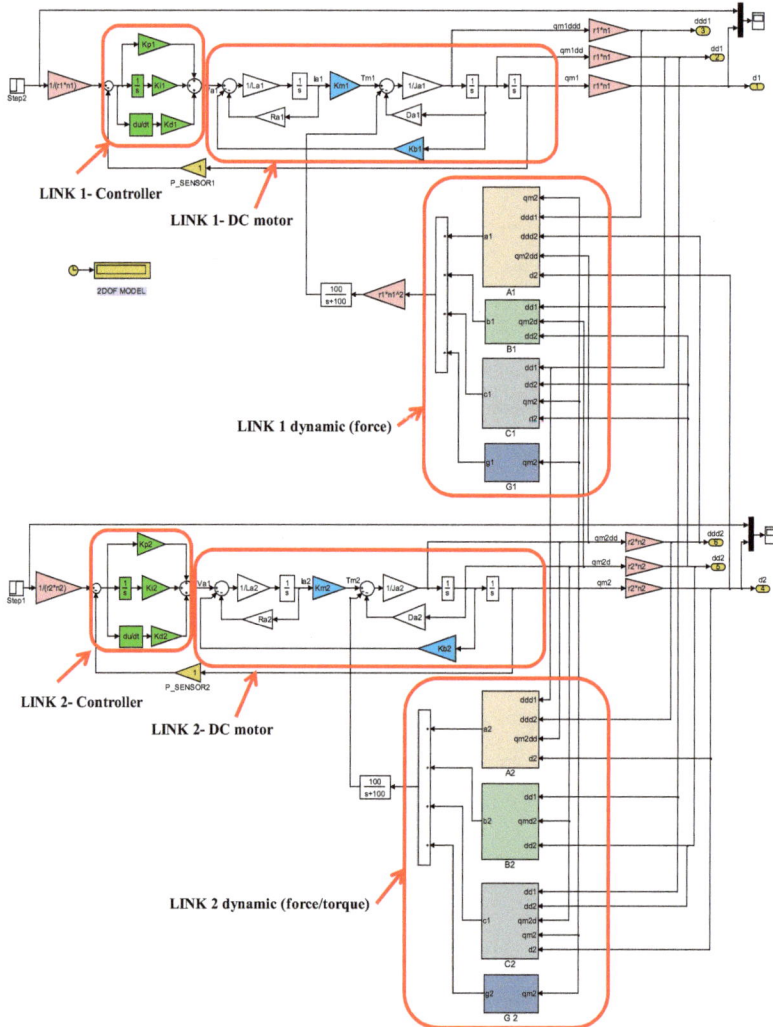

Figure 10: Block Diagram of the 2-Ttr Model.

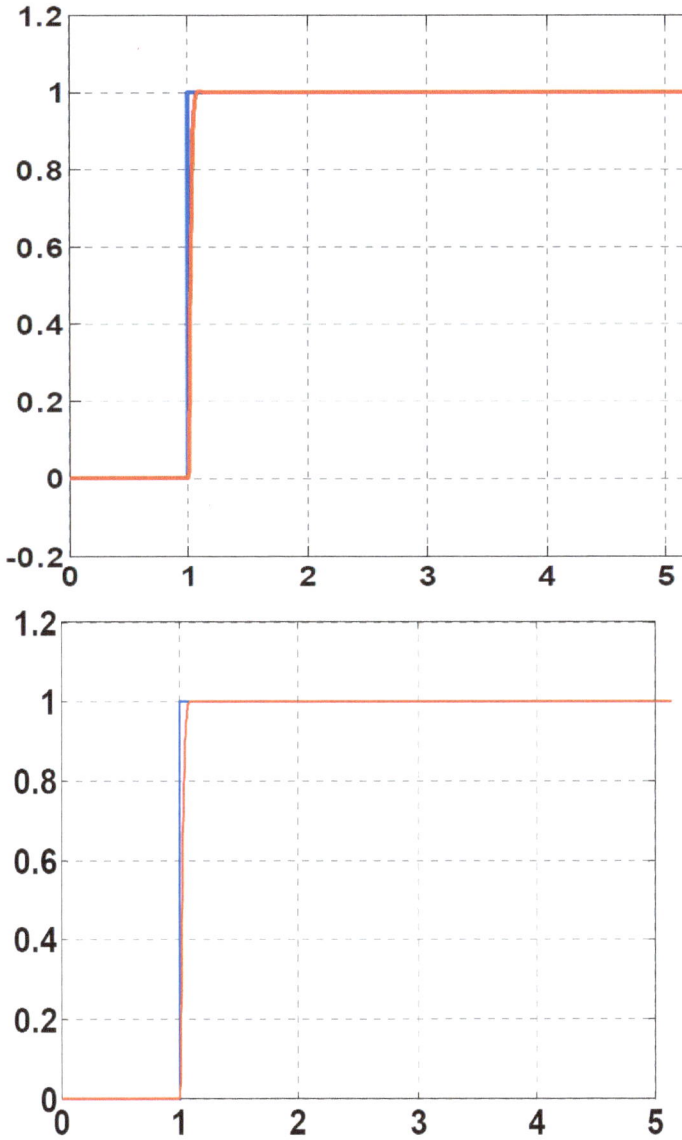

Figure 11: Time response of the Link 1 and Link 2 positions.

CONCLUSIONS

The reconfigurable modeling theory is used for the development of reconfigurable control systems that intelligently unify reconfiguration and manages the interaction of individual machine control systems within the RMS. An n-DOF

Global Kinematic Model (n-GKM) was previously developed for any combination of either rotational and/or translational types of joints.

An example the 2-GKM model is presented to demonstrate the methodology and application of the reconfigurable modeling theory. For the symbolic calculation of the 2-GKM dynamic equations, the recursive Newton-Euler algorithm is employed using the symbolic algebra package MAPLE 12. The dynamic model is named Global Dynamic Model (2-GDM). The significance of the 2-GDM is that it automatically generates each element of the inertia matrix A, Coriolis torque matrix B, centrifugal torque matrix C, and the gravity torque vector G, using Automatic Separation Method (ASM) and predefined reconfigurable parameters. These parameters are used to control the joint's positive directions and its type (rotational and/or translational). Instead of solving the dynamics of different kinematic structures, the 2-GDM can be used to auto-generate the solution by only defining the reconfigurable parameters. Using 2- GDM equations, a simple example of Ttr (Translation and translational/rotational) structure was used to demonstrate the model capability. Its kinematic, dynamic and control platform solutions are generated using previously defined reconfigurable parameters. The results proved the model's validity.

The reconfigurable modeling theory can be used for a current manufacturing system and for the future reconfigurable industry. The presented methodology can be applied to any automated machine, to develop reconfigurable modules and reconfigurable machines for the control processes.

ACKNOWLEDGEMENT

Declared none.

CONFLICT OF INTEREST

The author(s) confirm that this chapter content has no conflict of interest.

NOMENCLATURE

a_i : Link length of the common normal, $i = 1, 2, ..., n$

d_i Link offs*et al*ong previous z to the common normal, $i = 1, 2, ..., n$

θ_i : Joint angle about z, from old x to new x, $i = 1, 2, ..., n$

$^i(^0\omega_i)$: Angular velocity for rotational and translational joints, $i = 1, 2, ..., n$

$^i(^0V_i)$: Linear velocity for rotational and translational joints, $i = 1, 2, ..., n$

$^i(^0a_i)$: i^{th} Linear acceleration, $i = 1, 2, ..., n$

$^i(^0a_{ci})$: i^{th} Linear acceleration of the center of mass, $i = 1, 2, ..., n$

I_i : i^{th} Moment of inertia about center of mass of each link, $i = 1, 2, ..., n$

q : Vector of generalized joint coordinates

$\dot{q}$: Vector of joint velocities

$\ddot{q}$: Vector of joint acceleration

$^{i-1}P_i$: i^{th} Position matrix, $i = 1, 2, ..., n$

$^{i-1}R_i$: i^{th} Rotational matrix, $i = 1, 2, ..., n$

$(^{i-1}R_i)^T$: Transpose of all rotational matrices, $i = 1, 2, ..., n$

$^i(^0\alpha_i)$: i^{th} Angular acceleration, $i = 1, 2, ..., n$

θ_i : i^{th} Joint angle, $i = 1, 2, ..., n$

$^i(^0\omega_i)$: i^{th} Angular velocity, $i = 1, 2, ..., n$

$^{i-1}\dot{P}_i$: Linear velocity vector, $i = 1, 2, ..., n$

$^{i-1}\dot{\theta}_i$: Angular velocity vector, $i = 1, 2, ..., n$

$^{i-1}\ddot{P}_i$: Linear acceleration vector, $i = 1, 2, ..., n$

$^{i-1}\ddot{\theta}_i$: Angular acceleration vector, $i = 1, 2, ..., n$

^{i}g : Gravity vectors, $i = 1, 2, ..., n$

G_x, G_y, and G_z : Control parameters for the gravity vector

$^{i}(f_i^*)$: Joint forces

$^{i}(n_i^*)$: Joint torque

$^{6}(^{6}f_{Tool}) = 0$: End-effector force

$^{i}I_i$: Inertia tensor

T_m : Motor torque

K_m : Motor torque constant

I_a : Armature current

J_a : Armature inertia

R_a : Armature resistance

L_a : Armature inductance

T_L : Load torque

T_d : Disturbance torque

V_a : Armature voltage

V_b : Back Electromotive Force (BMF)

K_b : Back Electromotive Force constant

ω : Motor angular velocity

θ_m : Motor angular position

θ_L : Load angular position

T_L : Load torque

J_L : Load inertia

D_L : Load viscose damping

k_L : Load spring coefficient

n : Gear ratio

r_i : i^{th} Radius

N_L : Number of teeth of the output gear (load gear).

N_m : Number of teeth of the input gear (motor gear).

F : Translational link force

X : Translational link position

D_T : Translational link viscose damping

m : Translational link mass

α_i : Twist angle about common normal, from old z axis to new z axis, $i = 1, 2, ..., n$

R_i : Joints types control parameters $i = 1, 2, ..., n$

T_i : Joints types control parameters $i = 1, 2, ..., n$

T_{L2} : Link 2 torque

K_{Si} : Twist angles sinus control parameter, $i = 1, 2, ..., n$

K_{Ci} : Twist angles cosines control parameter, $i = 1, 2, ..., n$

A : Inertia matrix

B : Coriolis torque matrix

C : Centrifugal torque matrix

G : Gravity torque vector.

F_i - Link's force

a_{11} , a_{12} ,..., a_{1n} , ..., a_{nn} : Elements of the matrix A

b_{112} , b_{113} ,..., $b_{1(n-1)n}$, ..., $b_{n(n-1)n}$: Elements of the matrix B

c_{11} , c_{12} ,..., c_{1n} ,..., c_{nn} : Elements of the matrix C

g_1 , g_2 ,..., g_n : Elements of the matrix G

C , P , J_A , J_B : Variables used to determine geometry in Figs. **3** and **4**

C_x , C_y : Projections of point C

P_x , P_y : Projections of point P

config : Configuration parameter

m_i : Link one mass, $i = 1, 2, ..., n$

P_{Ci} : enter of link one, $i = 1, 2, ..., n$

G_x, G_y, G_z : Control parameters for the gravity vector

θ_{DHi} : Variable symbol of revolute joint, $i = 1, 2, ..., n$

d_{DHi} : Variable symbol of prismatic joint, $i = 1, 2, ..., n$

X, Y, Z : Cartesian Coordinate Frame

X_{i-1}^{ij} : Local Cartesian Coordinate Frames, .., $j = 1, 2, ..., n$, $k = 1, 2, ..., n - 1$

I_i : Inertia of links, $i = 1, 2, ..., n$

I_{xi}, I_{yi}, I_{zi} : Inertia components in X, Y, Z Cartesian frames, $i = 1, 2, ..., n$

REFERENCES

[1]   H. A. ElMaraghy, *Changeable and Reconfigurable Manufacturing Systems*, Springer-Verlag (Publisher), ISBN: 978-1-84882-066-1, 2008.

[2]   R.M. Shahin, W. H. ElMaraghy, E. M. ElBeheiry , "On nonlinearity control of CNC feed drives", *Trans. Can. Soc. Mech. Eng.* vol. 30, no. 1, pp. 63-80, 2005.

[3]   A. M. Djuric, *"Reconfigurable Kinematics, Dynamics and Control Process for Industrial Robots"*, PhD dissertation, University of Windsor, Canada, 2007.

[4]   A. M. Djuric and W. H. ElMaraghy, "Generalized Reconfigurable 6-Joint Robot Modeling", *Trans. CSME*, vol. 30, no. 4, pp. 533-565, 2006.

[5]   A. M. Djuric and W. H. ElMaraghy, "A Unified Reconfigurable Robots Jacobian", In: *2nd International Conference on Changeable, Agile, Reconfigurable and Virtual Production* (CARV), 2007, pp. 811-823.

[6]   A. M. Djuric and W. H. ElMaraghy, "Automatic Separation Method for Generation of Reconfigurable 6R Dynamic Equations", *Int. J. Manuf. Technol.,* vol. 46, no. 5-8, pp. 831-842, 2010.

[7]   A. M. Djuric and W. H. ElMaraghy, "Filtering Boundary Points of the Robot Workspace", In: *5nd International Conference on Digital Enterprise Technology*, Nantes: France, 2008

[8]   A. M. Djuric and W. H. ElMaraghy, (2008). "Unified Dynamic and Control Models for Reconfigurable Robots", In: *Changeable and Reconfigurable Manufacturing Systems*, Springer-Verlag (London) Ltd., 2008, pp. 147-161.

[9]   A. M. Djuric and W. H. ElMaraghy, "Comparison between Unified and Classical Modeling of Robotic Systems", In: 3*nd International Conference on Changeable, Agile, Reconfigurable and Virtual Production* (CARV), Munich, 2009.

[10] A. M. Djuric, R. Al Saidi and W. H. ElMaraghy, "Global Kinematic Model Generation for n-DOF Reconfigurable Machinery Structure", In: *6th IEEE Conference on Automation Science and Engineering*, (CASE 2010), Toronto, Canada, 2010.

[11] A. Djuric, R. Al Saidi, and W. ElMaraghy, "Dynamics Solution of n-DOF Global Machinery Model," *Robot. Comput.-Integr. Manuf.,*, submitted.

[12] B. Salem, M. Moll, and W-M. Shen, "SUPERBOT: A Deployable, Multi-Functional, and Modular Self-Reconfigurable Robotic Systems," In: *International Conference on Intelligent Robots an Systems*, (IEEE/RSJ), 2010, pp. 3636-3641.

[13] W-M. Shen, B. Salem, and Peter Will, "Hormones for Self-Reconfigurable Robotics," *Intell. Auton. Syst.,* vol. 6, pp. 918-925, 2000.

[14] S. Hirose, "Biologically Inspired Robots: Snake-Like Locomotors and Manipulators," *Oxford University Press*, New York, 1993.

[15] T. Fukuda and Y. Kawauchi, "Cellular Robotic System (CEBOT) as one of the realization of Self-Organizing Intelligent Universal Manipulator," In: *IEEE International Conference on Robotics and Automation*, 1990, pp. 662-667.

[16] G. S. Chirikjian, "Kinematics of a Metamorphic System," In: *IEEE International Conference on Robotics and Automation*, 1994, pp. 449-455.

[17] L. Kelmar and P. K. KHhosla, "Automatic Generation of Kinematics for a Reconfigurable Modular Manipulator System", *J. Robot. Syst.,* vol. 7, no. 4, pp. 599-619, Aug. 1990.

[18] M. Yim, D. G. Duff, and K. D. Roufas," POLYBOT: a Modular Reconfigurable Robot", In: *IEEE International Conference on Robotics and Automation, 2000.*

[19] F. Aagihili, K. Parsa, "A Reconfigurable Robot with Lockable Cylindrical Joints," *IEEE Trans. Robot.*, vol. 25, no. 4, August 2009.

[20] T. Strasser, M.N. Rooker and G. Ebenhofer, "Distributed Control Concept for a 6-DOF Reconfigurable Robot Arm", In*: Innovation production machines and systems: Fourth I*Proms Virtual International Conference*, 2008.

[21] Z. M. Bi, W.J. Ahang, I.-M. Chen and S.Y.T. Lang, "Automated Generation of the D-H Parameters for Configuration Design of Modular Manipulators," *Robot. Comput.-Integr. Manuf.,* vol. 23, pp. 553-562, 2007.

[22] Z. Li, W. Melek, C.M.Clark," Development and Characterization of a Modular and Reconfigurable Robot," In: *2nd International Conference on Changeable, Agile, Reconfigurable and Virtual Production* (CARV), 2007.

[23] S. Tabandeh, C. Clark , W. Melek, "Task-based Configuration Optimization of Modular and Reconfigurable Robots using a Multi-solution Inverse Kinematics Solver", ", In: *2nd International Conference on Changeable, Agile, Reconfigurable and Virtual Production* (CARV), 2007.

[24]   Spong M. W. and Vidyasagar M., *"Robot Dynamics and Control"*, J. Wiley and Son, New York, 1989.

[25]   T. Tarn, A. K. Bejczy, X. Yun, and Z. Li, "Effect of Motor Dynamics on Nonlinear Feedback Robot Arm Control", *IEEE Trans. Robot. Autom.,* vol. 7, no. 1, pp. 114-122, 1991.

Send Orders for Reprints to reprints@benthamscience.net

CHAPTER 6

Toward a Cognitive Assembly System

Li-Ming Ou and Xun Xu[*]

Department of Mechanical Engineering, University of Auckland, Auckland, New Zealand

Abstract This work is focused on a conceptual framework that is able to provide cognitive capabilities to an assembly environment. The framework provides assembly decision-making processes that consider assembly resources and design of a product. The proposed cognitive assembly system comprises of knowledge databases, assembly sequence generation, assembly resource planning and autonomous (self-learning) control capabilities.

Keywords Cognition, intelligent assembly, flexible assembly, human-robot cooperation, human-robot interaction, assembly systems, assembly sequence planning, self-learning, computer aided assembly planning, robot learning, robot planning.

INTRODUCTION

Today's market place is a very competitive arena. To gain an edge over competitor's product development, companies are offering end users ability to customize products. Rapid advancement of technology has also resulted in products that are more complex and customised. Complexity of products has brought about an increased number of components and manufacturing difficulties. Customised products have also increased the need for a more flexible production line, capable of producing different products. Current manufacturing systems are not suited for these manufacturing demands.

Cognitive assembly system is one of the key approaches to achieving flexible manufacturing. Such a system allows for a single production line to manufacture a wide variety of products, without compromising the advantages of a mass

*Address correspondence to Xun Xu: Department of Mechanical Engineering, University of Auckland, Auckland, New Zealand; Tel: +649 373 7599; Fax: +649 373 7479; E-mail: x.xu@auckland.ac.nz

production system. To achieve this type of flexibility and yet maintain the efficiency of an automated system, cognitive capabilities are often required. In an assembly system where robots work alongside human operators, robots need to be empowered with self-learning capabilities and cognitive functions. This entails tasks such as identifying and formulating interactions between robots and humans, modeling the interactions, and carrying out intelligent planning of manufacturing and assembly sequences.

ASSEMBLY SYSTEMS

The desire to perfect manufacturing processes has led to the development of many different types of manufacturing systems, each with its advantages and disadvantages. Manufacturing systems are selected by, or developed based on, the desired criteria, such as cost of implementation, type of production, volume of production, amount of workers available and product delivery time frame. Manufacturing systems can be classified into three major categories, one-off production, batch production and mass production systems. One of the most labour-intensive parts of a manufacturing process is assembly of products. Therefore, to increase flexibility and efficiency of the manufacturing process, assembly plans need to be developed.

1. A one-off production scheme is the basic method of producing products. This type of system is operated by highly trained craftsmen. A typical production using this scheme could be of a large scale product (*e.g.* an aircraft), production of custom designed products or products which are still at the development stage. This type of system is highly flexible and is capable of producing a variety of different products. However, the efficiency of such a production is low. One-off production systems usually have high costs due to labour-intense processes.

2. Batch production schemes are designed for medium-to-small volume productions. It is capable of producing different lines of products. This type of system is typically made up of several individual work-stations, each specialised at specific tasks that are mostly carried out

by human experts. Due to the reliance on human experts, this type of production requires a high level of training for producing new types of products. Work stations are most predominantly made up of manual operations with minimal amount of automation introduced.

3. Mass production systems are designed to deliver a large volume of production at an efficient level. However, this efficiency often comes at the cost of flexibility in production. This type of production is usually planned with care and designed for continuous production of a single type of identical products. Development of mass production schemes has led to a high degree of automation with an aim of reducing labour cost and achieving higher quality control. Increase in automation has resulted in use of fixed tooling, specialised machinery and delivery systems. The objective of this type of system is to produce a large volume of quality products with the least amount of time and cost.

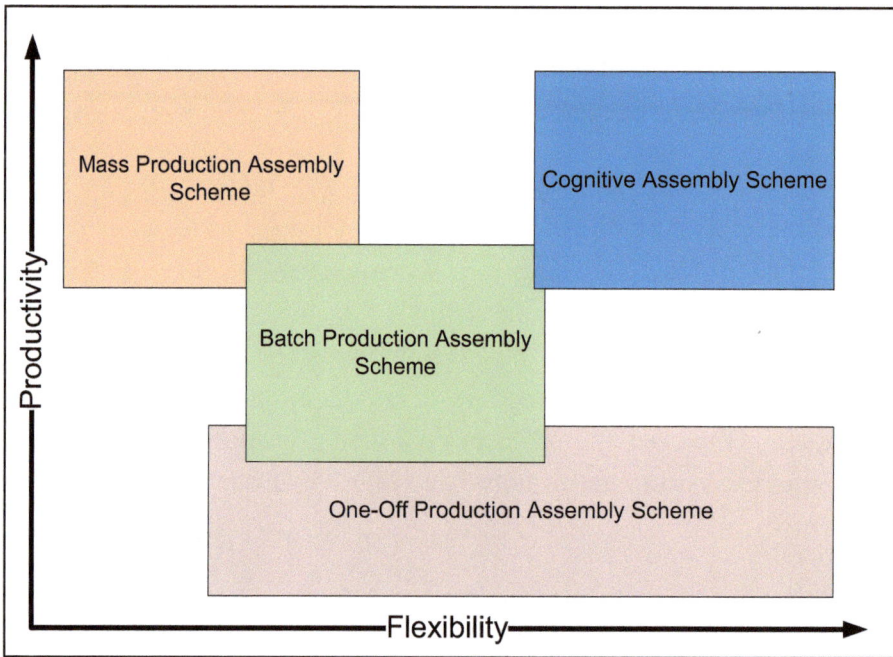

Figure 1: Different classifications of assembly schemes in manufacturing systems with the introduction of the cognitive assembly scheme.

Assembly processes in a one-off production scheme are carried out manually. Typically, in a one-off scheme the work-in-progress (WIP) is stationary at one location during assembly. Work is carried out around the WIP, tools and resources are transported to the WIP. However, transporting equipment and resources to a single location when carrying out multiple assemblies can be inefficient and time-consuming. Therefore, assemblies began to be separated into smaller subassemblies. Subassemblies are delegated to different assembly workers and performed simultaneously. This assembly method can be seen in batch production schemes. In batch production, WIP is no longer remained in a single location. It moves to different workstations where subassemblies are carried out. Equipment and resources required for each sub assembly is also kept in the workstations reducing the need for transportation. This increases efficiency and reduced worker's work load. Assembly is further developed by employing specialised machines, which are able to carry out specific assembly tasks. With the use of machines and dividing a large complex assembly into smaller subassemblies, assembly lines are developed. These assembly lines have been introduced into mass production schemes and certain batch production schemes. Assembly processes have been further developed through use of automation. Fully automated systems have been achieved by simplifying product designs. Fig. **1** shows how the cognitive assembly scheme is compared to the three above mentioned assembly schemes in terms of productivity and flexibility. Table **1** highlights the specific advantages and disadvantages of all three systems, which can affect the development of a cognitive assembly system.

Table 1: Advantages and disadvantages of the three main types of production schemes

	Advantages	Disadvantages
Assembly in Mass Production	• High efficiency in production cost and production rate • High level of automation • Available for continuous production • Automation reduced human operator contact with hazardous situations	• No flexibility in product variety • Requires an experienced programming team • Workers have no interactions with robots • No automated assembly sequence generators, still relies on human experts • Automation is achieved by specifically designed components • Specialised machines only operate for specific components

Table 1: contd…

Assembly in Batch Production	• Able to produce different products between batches with some flexibility	• Requires worker training for production of different products • Typically utilise manual workstations • Rely human operators • No automated assembly sequence generator • No direct robot collaboration • Lower efficiency than mass production
Assembly in One-off Production	• Highly flexible in the type of product that can be assembled	• Required highly trained and experienced workers • Typically utilise manual workstations • No automated assembly sequence generator • No direct robot collaboration • Low productivity

For a robot-assisted assembly scheme to be cognitive, effective, efficient and capable of providing flexibility, it needs to meet the following requirements.

- Be able to achieve high production flexibility in a variety of products that are assembled on a single production line.

- Ability to prompt human workers on-demand with product information and instructions, thus reducing the amount of worker training.

- Ability to monitor human worker's process and decisions and provide optimum working environments.

- Ability to carry out direct human-robot interactions, allowing for tasks to be carried out simultaneously.

- Allow the workers to have direct control of their robot counterparts to ensure a smooth collaboration.

- Ability to plan and generate assembly plans, taking into account available resources.

- Ability to self-adapt and learn from the collected data. Data from the tasks carried out during operations.

COGNITION IN MANUFACTURING

To achieve an environment where a human worker and robot/intelligent equipment are operating in synergy, it is essential to understand the psychology of human workers. This understanding is essential for mimicking a human worker's behaviour and allowing effective human-robot collaborations. Different models have been developed for human's thinking process based on various cognitive processes. Cognitive processes can be broken down into two basic types, pattern recognition and the ability to store and retrieve data. It has been found that there are three different mechanisms that an operator recognises patterns;

Template-Matching

This is a method of recognising patterns by the use of a stored "template" [1]. This is only useful in circumstances where shapes do not change with time. This theory has two major faults. Firstly, a change in size can cause template-matching to fail. Secondly, rotation of the image may also cause template-matching to fail. This method was mainly developed for exact matching and theory work.

Prototype-Matching

This method is similar to template-matching, except that instead of storing a template a prototype is stored. "Prototype is idealised/abstract patterns" [1]. This is different from a template as it is a generalised pattern which can be used to recognise similar patterns and an exact match is not required.

Feature-Based Approach

This theory works by storing a set of features specific to a pattern. The features of an image can be used to match the features stored in the memory to identify the pattern.

A typical task carried out during an assembly operation is handling multiple components. With an increased flexibility, there is a need to differentiate between different components. Therefore, the ability of robots/machines to recognise patterns in a similar manner to human operators is one of the key requirements for achieving a cognitive assembly system. Pattern recognition can also provide a means for assembly sequence generators to recognise similarities among products and create assembly plans using an existing plan. This way, the amount of learning will be reduced when new product assembly sequences are introduced.

In a human cognitive system, there are two types of memory storage, short term memory (STM) and long term memory (LTM). Models have been developed to understand the framework of human memory system based on different theories. There are three main theories.

Atkinson-Shiffrin's Model

This model structure is based on the sensory information that one receives [1, 2]. These senses include sight, sound, touch and smell. This information is stored in sensory registers. These registers store the most recent events but decay rapidly to allow storage of new data. The memories stored in the sensory registers are duplicated in the short term memory and the memories in the STM are then partially copied into the long term memory. This theory can be compared to a computer memory system. Sensory registers can be said to be the input of the computer. STM can be compared to the cache of a computer system and LTM is the RAM. The human process unit (the brain) is the CPU. However, many researchers questioned this theory in that they believe LTM and STM are not completely separate and some believe that the sensory registers can interact directly with the LTM [1].

Tulving's Model

This model is based on three types of memory that are collected from the environment and problem-solving. The three types are episodic, semantics and procedural [3]. The episodic memories are used to the event currently occurring. The semantics memory is the knowledge database. The procedural memory is used for executing a task based on the previously stored memory.

Parallel Distributed Processing Model

This model represents memory as network singular memories that can store information simultaneously. This model allows machine learning, but it does not differentiate STM from LTM similar to that of a human's memory system. The information stored can be accessed in parallel for executing tasks [4].

Memory systems allow operators to build up an extensive knowledge database about assemblies and the required motions to be performed in order to complete a task. A similar system is required for implementing a cognitive assembly system.

The main drive for implementing a cognitive factory is to be able to take advantage of both human operators and industrial machines. The ability with which machines operate endlessly without rest yet maintain quality and a high-rate of production is a valuable attribute in manufacturing. Human workers have the ability to think and solve problems on the fly and possess extensive dexterity, all of which are equally essential for manufacturing [5]. A system in which humans and machines work in harmony and collectively shall provide superior capabilities. Such a system is much needed with the current industry shifting from a seller's market to a buyer's market, where customers are have more and more say in their products. This change is already very evident in the computer market where computers are made to order, with customers being able to pick and choose the components of a computer.

The increasing requirement of a more flexible factory environment has led to development of a cognitive factory. From current research, cognitive factory can be divided into five steps [5].

1. Develop a cognitive control system similar to a human cognitive system to enable human-machine collaboration.

2. Build a dynamic knowledge database that can update in real-time to take account of the working environment and any changes that occur.

3. Generate machine control and trajectories automatically in real-time.

4. Develop an assembly planning system that can allocate, analyse the product and create assembly sequences given the design of the assembly. The assembly plan should incorporate both human and machines.

5. Develop a system that can execute and maintain the production with consideration of the human partner(s) safety.

To model human cognitive capabilities, two criteria need to be achieved. The first criterion is the ability to solve problems systematically. The second is the ability to calculate and handle a large dynamic database [1]. One of the main weaknesses with all the current models is the inability to cope with representing a parallel and distributed human way of thinking. There are some research methods reported in the literature in modeling human cognitive systems to enable parallel capabilities. These methods include Petri Nets, Fuzzy logic, neural networks, Genetic algorithms, multi-agents and a combination of different techniques.

ASSEMBLY SEQUENCE PLANNING

Due to the complexity of mechanical products, there is always more than one way of assembling them. In the past, assembly planning has involved planning of materials, scheduling and sequences. These planning tasks were done manually by an experienced operator, which is a time-consuming process. With an increase in product complexity these manual processes become inefficient. Therefore, research has been carried out for generating assembly sequences automatically.

In early stages of sequence generation development, the focus is on knowledge-based approaches through the use of questions, graphs and tables to generate different sequences. Bourjault developed a 'Liaison model' which is made up of a series of questions about the assembly and the components [6]. Based on the answers to the questions, a knowledge base was created that can aid sequence generation. Due to the nature of the questions, the number of questions expanded with the increase in the number of components. For some complex assemblies, the number of questions can become impractical to answer. De Fazio and Whitney later improved the model by streamlining the questions and reducing the number of questions [7].

The 'Liaison model' however was never able to completely represent the sequence generation. Homan de Mello and Sanderson developed the AND/OR graphs, which allowed a more complete representation of an assembly sequence

[8]. The AND/OR graph uses a series of nodes and arcs to represent all the available assembly sequences. The nodes represent the components and subassemblies of assembly. The arcs illustrate the different approaches that can be taken to assemble the product. The AND relations show the subassemblies required to assemble the product. The OR relations show all the different combinations of the components and subassemblies that are available. There is a disadvantage in the ability to represent the sequences completely as the graph would become very large and hard to handle with the increase in the number of components.

Ko and Lee developed a 'Part-ordering method' which explores the mating conditions and interactions between the different components [9]. This method ensured that there are no collisions between parts during the assembly process. Ben-Arieh and Kramer developed a model that generates the assembly sequences by considering the precedence relationships and the different combinations of subassemblies [10]. Huang and Lee, Dini and Santochi, developed mathematical models to represent the information required for the sequence generation which includes an assembly structure, component precedence relationship and recourse constraints [11, 12]. However, this research expected that the knowledge of the assembly is already available in the form of human experts. The information required by these systems is also more than what CAD models can provide.

Tiam-Hock Eng *et al.* developed a 'Kinematic pair liaison diagram' that studies the degree of freedom of the components using matrices based on the features of the components [13]. This model uses geometric reasoning to distinguish the relations between the parts. This model assumes that the component precedence knowledge is already available. Eng *et al.* developed the system to allow possibility to use with CAD models. Kaufman *et al.* created a system that optimised assembly sequence plans from a CAD model for use on a robot [14]. However, this was limited to a single specific class of assembly. Mazouz *et al.* incorporated artificial intelligence and the knowledge based approach to assembly sequence generation [15]. Gu and Yan developed a featured-based system that uses CAD systems [16]. A prerequisite is the knowledge of the assembly description. The system does not generate all the sequences but is used for optimising the assembly sequence. De Lit *et al.* uses Genetic Algorithm to generate an assembly tree for generating assembly sequences [17]. This requires a starting assembly tree which is generated randomly.

Zha and Du used STEP for a feature-based model though it requires prior knowledge of the assembly relation [18]. Gottipolu and Ghosh developed a system that can use a CAD solid modeler to directly create an assembly sequence [19]. The system is contact-based and utilises two matrices to illustrate the presence or absence of contacts in generating feasible sequences. Similarly, Maziero *et al.* developed a feature-based assembly sequence generator by examining the surfaces of the components for the presence or absence of contacts [20]. The system is able to automatically extract information from a group of input components. However, the system is restricted only to cylindrical types of components. Jayaram *et al.* developed a system which utilizes a CAD system and a supplementary system to generate an assembly plan [21]. The system developed incorporates a CAD system and a visualisation tool that supports reorganisation of the components that better represents the assembly sequence while preserving and updating all the information from the CAD model [21].

The research carried out by Chaudron *et al.* generates an assembly sequence using the combination of AND/OR graphs and CAD software [22]. The research suggests generating an assembly sequence by grouping the components to allow the worker to decide a specific sequence for that group of components. This is done by considering the structural connection between the components. Su developed a 'Geometric constraints analysis' approach which is integrated with a CAD system [23]. The system derives the precedence relations automatically. This system still requires human-computer interactions for certain constraint information.

The majority of the current research on assembly planning has focused on sequence generation without considering implementation, or with no consideration of an actual assembly system. The current assembly system has the potential to be further enhanced to provide more flexibility, through the collaboration between human and robots or intelligent machines. To enable this collaboration, cognitive capabilities need to be included in the system. Cognitive capabilities can provide a link between the assembly sequence generation and the implementation with consideration of resources and human-robot collaboration.

EXECUTION OF ASSEMBLY PLANS

Robot Programming and Learning

Researchers have worked on giving robot the ability to perform self-learning. The user-orientated programming methods provide a good support for a human-robot assembly system.

Lopez-Juarez *et al.* carried out research on integration of learning capabilities with intelligent robotic agents for assembly [24]. The cognitive architecture for the robotic agent is a development based on an adaptive resonance theory as a knowledge-based artificial neural network (ART). It is used to imitate human cognitive and learning processes. The cognitive architecture is used to accomplish assembly tasks. The researcher also used a Fuzzy ARTMAP network for classification and recognition of simple components. This work is limited to simple parts and only operates on a top-down view system. Voyles and Khosla used a multi-agent system for developing a Gesture-based robot programming method [25]. The method relies on previously acquired robot skills in the form of a list of skills that allow the robot to perform assembly and other possible tasks. Robot programming is carried out by the use of data gloves and fingertip covers together with fuzzy rules to recognize the actions of a demonstrator. The work is currently limited to the peg-in-hole type of operations. Liu and Nakamura used skill primitives for robot control [26]. This research does not use human demonstrators to create programs. It creates skills automatically from an assembly plan for a two-finger precision manipulator. The system is based on solid models using the assembly and mate trees within the models. Skoglund *et al.* explored the idea of gesture programming and a similar approach has been followed using data gloves for the demonstrator [27]. However, this approach uses both a fuzzy modeling and a next-state planner which allows the automatic acquisition of skills equivalent to human gestures and the system to adapt previously stored skill primitives to new skills. Wang and de Silva developed a system for "multi-robot coordination" in a dynamic environment [28]. This research used the theory of object transportation which requires learning capabilities to avoid collisions.

Reinforcement learning in the form of sequential Q-learning and Genetic Algorithm has been used to give self-learning capabilities. Eraslan and Kurt developed a method for evaluating the cognitive performance factors for a flexible manufacturing system using fuzzy theory [29]. This allows workers to better monitor an assembly system. Bi *et al.* explored the concept of "adaptive robotic system" for industrial robotics systems [30]. An adaptive robotic system can be separated into two main approaches, adjustable systems and modular systems. Bi *et al.* has developed a hybrid system called "adaptive robotic system architecture" [30], combining the advantages of both approaches. The aim was to create an adaptable system that is able to morph into multiple variants to suit different situations and also to meet the manufacturing needs. This system utilizes two types of matrices for adaptive capability - the incident matrix in graph theory and the object indirect matrix.

Human-Robot Cooperation

It has been shown in practice that there is an obvious advantage when the human and robot can collaborate on tasks such as handling heavy equipment or material, pick and place operations and programming robots *via* direct contact.

Ogorodnikova proposed an application of human-robot collaboration for a disassembly cell [31]. This research studies the human factors in a collaboration scheme. The human factors that need to be monitored in order to achieve an efficient and safe collaboration include ergonomics, human workload, human proficiency, auditory presentation, visual focus and the workspace allocation. Lenz *et al.* proposed a human robot collaboration system that allows human and robots to work in the same domain [32]. In a traditional human robot collaboration system, the robot acts as a slave to the operator and collaboration occurs in different domains. For joint actions, task sharing and coordination to occur in the system, the robot is required to anticipate an operator's actions by the use of knowledge database and decision making processes. A concept of such a system has been demonstrated by Lenz *et al.* in a LEGO building operation. Wojtara *et al.* developed a system for human robot collaboration in handling large components [33]. Most of industrial robots are not equipped with sensory to avoid collisions and obstacles. Collaboration was achieved by first studying human-

human collaboration behaviour and based on the acquired data, a system which responds to human behaviour can then be developed. This system is limited by the dependency on human workers as instructions and decisions are made entirely by them. Mayer *et al.* studied direct interactions between human operators and robots [34]. The research was applied in a simple LEGO building task which identified the importance of processing human-working behaviour and the ability to automatically comply with the behaviour. Kim *et al.* developed a human-machine collaboration model based on the finite state automaton [35]. Like other researchers, they identified the need to model human's behaviour in collaborations. Kruger *et al.* studied the cooperation with machines in assembly lines and classified the assembly system into three categories: manual, cooperative and automatic systems [36]. The research addresses the issue of work-place sharing and the interfaces between human and intelligent machines. The interfaces concerned are visual, gesture, speech and physical.

Safety

The risk around the interaction of a robot and a human operator is considered as high in current systems [37, 38]. A typical commercial industrial robot does not have sensors or controllers built in to detect any collisions. The limited safety features available commercially include light gates, physical barriers and emergency stops or switches. Therefore, for a human and robot to collaborate in a more unified environment a more advanced safety system is required.

The current safety requirement for a robot to work in the manufacturing environment is that of complete separation between the robot and the operator when the robot is active. This entails employment of the safety equipment detailed above. To develop a new safety standard, collisions of the robot need to be studied. Typically, there are three main types of robot collisions [37], product-related collisions, collisions as results of control errors and collisions with a human operator.

There has also been development in giving the robot a more advanced sensory system for collision detection. Research on capacitance sensors for use on robot arms has been carried out. The sensors act as artificial skin, which can detect collisions with objects such as human operators [38]. Wrist-mounted laser scanner has also been used to prevent collisions. Another approach is to use cameras to provide depth and positional information and together with image processing software to detect potential collisions. A robot can be monitored to varying degrees depending on the state of the cooperation. The state of cooperation can be separated into four categories, free motion, guided motion, gross motion and fine motion. Monitoring levels increase respectively. However, there is a major criterion that needs to be achieved in all four states. That is the need to avoid unnecessary collisions with anything at all times. The work carried out by Henrich *et al.* uses two main principles when applying the safety precautions [37]. They are adapted area of surveillance and distance controlled velocity.

The area of surveillance can be adapted to the state of collaboration. For example, when a robot is being guided by an operator, the focus of surveillance is on parts of the robot that is not in contact with the operator as those parts are more likely to cause harm to human. Distance-controlled velocity controls the speed of the robot when it approaches an obstacle which may be a work piece, obstruction or a human body part [37]. This principle is sensible also because it will cause less disruption to the operator when the robot arm approaches slowly since it is a basic human instinct to be aware of objects approaching at a high speed. This will cause a reduction in concentration on the task at hand.

DEVELOPMENT OF A COGNITIVE ASSEMBLY ENVIRONMENT

To develop a cognitive environment for an assembly system, further research is required and there is also a need to integrate different aspects of an assembly system. Some of the key research areas include a knowledge database, assembly sequence planning and resource management and autonomous control capabilities.

Knowledge Database

In assembly, there is a vast amount of knowledge to be utilised, from the design stage to the end product. The knowledge involved in carrying out an assembly task includes product knowledge, resource knowledge, scheduling knowledge and knowledge on general assembly tasks. Product knowledge is the knowledge about individual products and their functions which will aid the assembly process. Such knowledge may be used to determine if the assembly can be separated into multiple subassemblies and the knowledge of the materials from which the product is manufactured. Product knowledge also includes knowledge of the components precedence which is required for generating assembly sequences. Resource knowledge includes (a) the shop-floor resources, (b) machines and robots available together with their respective capabilities in performing assembly tasks, (c) human resources and (d) the location of raw materials and transport systems. Scheduling knowledge includes deadlines, volume and the sequence products need to be completed. The knowledge on general assembly task includes manual tasks required to complete an assembly such as fastening bolts and knowledge on best method of handling components of similar forms. For realising a cognitive assembly system, knowledge bases need to be integrated into a dynamic structure. During an assembly, the control system may be required to access data on the machines, the state of the assembly line and the data about the product. Through this integrated knowledge database, the assembly system will have the capability of being aware of the product requirements, responding by communicating with suitable resources and devising a plan for both human and machines.

Assembly Sequence Planning and Resource Management

There has been little effort made to integrate assembly planning with assembly resources available. The assembly planner's current focus is on creating an optimised assembly sequence based on cost and time. This does not consider the resources available in the factory that the plan is generated for. To integrate sequence planning and assembly resource management, cognitive capabilities are required. This system will need to be able to make changes to the assembly plan dynamically depending on the assembly sequence used by a human operator. An operator could for any reason during the assembly decide to change the assembly sequence. Therefore, the assembly sequence generator should be able to identify the chosen assembly sequence and adapt to match the chosen sequence. The assembly planning system should also be a robust and diverse system in that it is able to (a) identify the resource availability such as the robots available that can carry out the task and (b) select the most suited resource to carry out particular tasks in assembly. It should also be able to identify which task is most suited for human and which is most suited for robot to perform, so as to develop a plan of execution based on the respective strengths.

Autonomous Control Capabilities

A cognitive assembly system requires comprehensive autonomous control capabilities, which allow for direct human-robot collaboration and help achieve the same high efficiency as that of a mass production system. The required control capabilities are to do with material transport systems, robot trajectories, machine parameters and identification of work-in-progress. For a cognitive assembly system to operate efficiently, the control needs to be generated from assembly plans and be able to update dynamically in production. The system also requires new safety measures in place for direct human-robot collaborations. These measures include sensors and interactive monitoring devices between both human and robot.

The Framework

A framework has been proposed for a cognitive assembly system. The system takes a CAD model as input from which assembly sequences are generated. An optimized solution is obtained based on cost, resource, assembly time and effectiveness of human-robot collaborations. The cognitive system is part of a cognitive factory scheme that can be split into two components, (a) assembly decision-making and (b) self-learning (Fig. **2**).

The assembly-decision-making component provides capabilities for generating and assigning assembly sequences to human and robots. To generate and represent all the possible assembly sequences of a complex mechanical assembly, AND/OR graphs are used. Genetic algorithms are used to optimize assembly sequences. To enable collaborations between human and robots, robot "skill primitives" method is used. The system is also able to convert assembly sequences into robot codes and human instructions. An important capability of the assembly-decision-making component is its ability to distinguish and assign tasks that are suitable for human operators or robots to perform alone, or suited for human-robot collaboration.

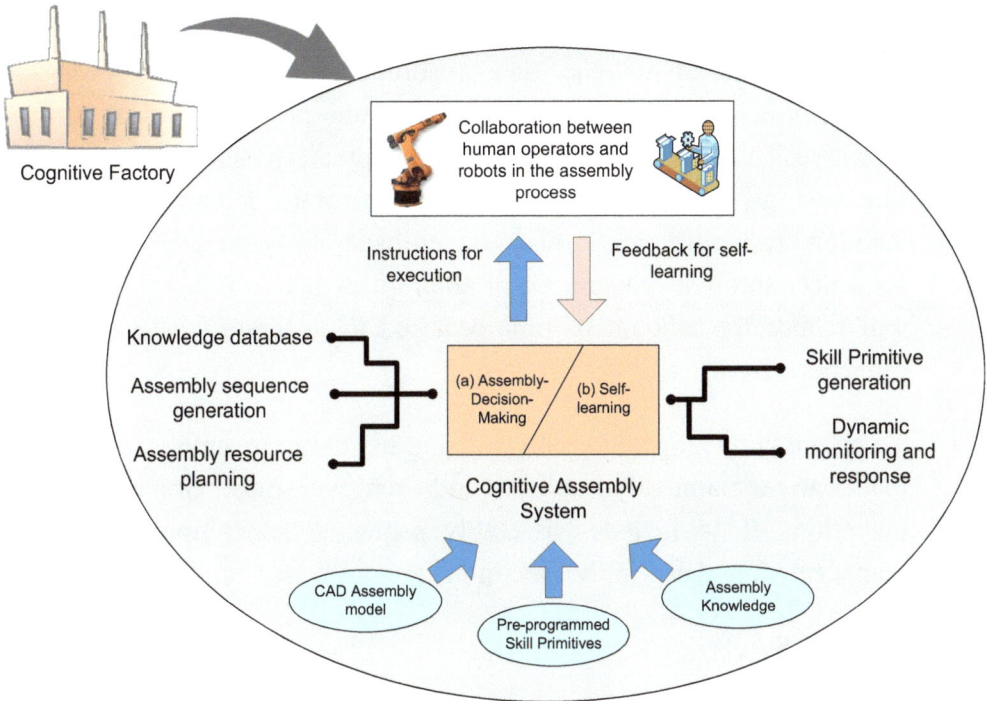

Figure 2: Cognitive assembly system conceptual framework.

The neural network approach is used to enable self-learning and adapting capabilities. This enables the system to generate new "skill primitives" based on the current skills and knowledge recorded during the assembly operations. Self-learning capabilities enable the system to react to the environment dynamically. During assembly, an operator may choose a different method of completing the task, such as changing the assembly sequence or using a different tool. Therefore, the system will react to changes and provide assistance accordingly.

The proposed framework has four objectives:

- To integrate assembly sequence generation with assembly resources so as to create resource specific assembly plans. This feature allows the generated assembly sequence to be relevant and practical for a particular factory and shop-floor resources.

- To enable direct human-robot collaboration on assembly tasks. Providing the system with robust decision-making capabilities.

- To give self-learning capability for generating new robot "skill primitives" based on the data recorded during assembly. This capability helps reduce the amount of off-line programming of "skill primitives" and allow robots to react to unpredicted situations in real-time and give the system a flexible and robust feature. The self-learning capability may also be utilised to generate assembly sequences through recognition of similarities between products. This will reduce the amount of time required to generate new assembly sequences.

- To give a knowledge database covering the entire assembly spectrum that can dynamically update and retrieve information during operations. This includes assembly sequence generation, resource management and assembly execution.

CONCLUSIONS

The existing research work on assembly sequence and assembly execution has identified key characteristics in achieving a cognitive assembly environment. These characteristics are:

- Automatic sequence generation that may be achieved with a combination of graphs (*e.g.* AND/OR graphs) and artificial intelligence (*e.g.* genetic algorithm).

- A "skill-primitive" approach to programming the robot.

- Self-learning based technologies such as artificial neural network, fuzzy modeling, genetic algorithm and reinforcement learning.

- Safety conscious human-robot collaboration.

- Task sharing and human operator assisting capabilities.

Based on the key characteristics identified, a conceptual framework is proposed that is able to provide cognitive capabilities to an assembly environment. The framework provides assembly decision-making processes that consider assembly resources and design of a product. Self-learning capabilities allow the system to dynamically adapt to the assembly environment. This framework comprises of knowledge databases, assembly sequence generation, assembly resource planning and autonomous (self-learning) control capabilities.

ACKNOWLEDGEMENT

Declared none.

CONFLICT OF INTEREST

The author(s) confirm that this chapter content has no conflict of interest.

REFERENCES

[1]   A. Konar, and L. Jain, "The Psychological Basis of Cognitive Modeling," In: *Cognitive Engineering*. London: Springer, 2005, pp. 1-38.

[2]   J.G.W. Raaijmakers, and R.M. Shiffrin, "Models for Recall and Recognition," *Annu. Rev. Psychol.*, vol. 43, no. 1, 205-234, 1992.

[3]   E. Tulving, "Multiple memory systems and consciousness," *Human Neurobiol., v*ol. 6, 67-80, 1987.

[4]   D.E. Rumelhart, and J.L. Mcclelland, *Parallel distributed processing: explorations in the microstructure of cognition. Volume 1. Foundations*, MIT Press: Cambridge, MA, 1986.

[5]   M.F. Zah, M. Beetz, K. Shea, G. Reinhart, K. Bender, C. Lau, M. Ostgathe, W. Vogl, M. Wiesbeck, M. Engelhard, C. Ertelt, T. Ruhr, M.Friedrich, and S. Herle, "The Cognitive Factory," In: *Changeable and Reconfigurable Manufacturing Systems*, H.A. ElMaraghy, Ed., Springer: London, 2002, pp. 355-371.

[6]   Bourjault, A., "*Contribution a une approache methodologique de l'assemblage automatise: elaboration automatique des sequence s operatories,*" PhD thesis, L' Universite de Franche-Comte, France, 1984.

[7]   T. De Fazio, and D. Whitney, "Simplified generation of all mechanical assembly sequences," *Robot. Automat.*, vol. 3, no. 6, 640-658, 1987.

[8]   L.S. Homem de Mello, and A.C. Sanderson, "A correct and complete algorithm for the generation of mechanical assembly sequences," *Robot. Automat.*, vol. 7, no. 2, 228-240, 1991.

[9]   H. Ko, and K. Lee, "Automatic assembling procedure generation from mating conditions," *Comput.-Aided Des.*, vol. 19, no. 1, 3-10, 1987.

[10]   D. Ben-Arieh, and B. Kramer, "Computer-aided process planning for assembly: generation of assembly operations sequence," *Int. J. Prod. Res.*, vol. 32, no. 3, 643-656, 1994.

[11]   Y.F. Huang, and C.S.G. Lee, "A framework of knowledge-based assembly planning," In: *Proc. IEEE International Conference,* 1991, pp. 599-604.

[12]   G. Dini, and M. Santochi, "Automated Sequencing and Subassembly Detection in Assembly Planning," *CIRP Ann. - Manuf. Technol.,* vol. 41, no. 1, 1-4, 1992

[13]   T.H. Eng, Z.K. Ling, W. Olson, and C. McLean, "Feature-based assembly modeling and sequence generation," *Comput. Ind. Eng.,* vol. 36, no. 1, 17-33, 1999.

[14]   S.G. Kaufman, R.H. Wilson, R.E. Jones, T.L. Calton, and A.L. Arlo, "The Archimedes 2 mechanical assembly planning system," In: *Proc. 13th IEEE International Conference on Robotics and Automation,* Minneapolis: USA, 1996, pp 3361-3368.

[15]   A.K. Mazouz, A. Souilah, and M. Talbi, "Design of an expert system for generating optimal assembly sequences," *Comput.-Aided Eng. J.,* vol. 8, no. 6, 255-259, 1991.

[16]   P. Gu, and X. Yan, "CAD-directed automatic assembly sequence planning," *Int. J. Prod. Res.,* vol. 33, no. 11, 3069-3100, 1995.

[17]   P. De Lit, P. Latinne, B. Rekiek, and A. Delchambre, "Assembly planning with an ordering genetic algorithm," *Int. J. Prod. Res.,* vol. 39, no. 16, 3623-3640, 2001.

[18]   X.F. Zha, and H. Du, "A PDES/STEP-based model and system for concurrent integrated design and assembly planning," *Comput.-Aided Des.,* vol. 34, no. 14, 1087-1110, 2002.

[19]   R.B. Gottipolu, and K. Ghosh, "A simplified and efficient representation for evaluation and selection of assembly sequences," *Comput. Ind.,* vol. 50, no. 3, 251-264, 2003.

[20]   N.L. Maziero, J.C.E. Ferreira, and F.S. Pacheco, "A method for the automatic identification of contacts in assemblies of cylindrical parts," *J. Braz. Soc. Mech. Sci. Eng.,* vol. 26, no. 3, 297-307, 2004.

[21]   U. Jayaram, Y. Kim, S. Jayaram, V.K. Jandhyala, and T. Mitsui, "Reorganizing CAD Assembly Models (as-Designed) for Manufacturing Simulations and Planning (as-Built)," *J. Comput. Inf. Sci. Eng.,* vol. 4, no. 2, 98-108, 2004.

[22]   V. Chaudron, P. Martin, and X. Godot. "Assembly sequences: planning and simulating assembly operations," In: Proc. *IEEE International Symposium on Assembly and Task Planning.* (ISATP 2005), 2005, pp. 156-161.

[23]   Q. Su, "Computer aided geometric feasible assembly sequence planning and optimizing," *Int. J. Adv. Manuf. Technol.,* vol. 33, no. 1, 48-57, 2007.

[24]   I. Lopez-Juarez, J. Corona-Castuera, M. Peña-Cabrera, and K. Ordaz-Hernandez, "On the design of intelligent robotic agents for assembly," *Inf. Sci.,* vol. 171, no. 4, 377-402, 2005

[25]   R.M. Voyles, and P.K. Khosla, "Multi-agent system for programming robots by human demonstration," *Integr. Comput.-Aided Eng.,* vol. 8, no. 1, 59-67, 2001.

[26]   Z. Liu, and T. Nakamura, "Combination of robot control and assembly planning for a precision manipulator," *Int. J. Adv. Manuf. Technol.,* vol. 31, no. 7-8, 797-804, 2007.

[27]   A. Skoglund, B. Iliev, and R. Palm, "Programming-by-Demonstration of reaching motions-A next-state-planner approach," *Robot. Auton. Syst.,* vol. 58, no. 5, 607-621, 2010.

[28]   Y. Wang, and C.W. de Silva, "A machine-learning approach to multi-robot coordination," *Eng. Appl. Artif. Intell.,* vol. 21, no. 3, 470-484, 2008.

[29]   E. Eraslan, and M. Kurt, "Fuzzy multi-criteria analysis approach for the evaluation and classification of cognitive performance factors in flexible manufacturing systems," *Int. J. Prod. Res.,* vol. 45, no. 5, 1101-1118, 2007.

[30]   Z.M. Bi, Y. Lin, and W.J. Zhang, "The general architecture of adaptive robotic systems for manufacturing applications," *Robot. Comput.-Integr. Manuf.,* Article in press.

[31]   O. Ogorodnikova, "Human Weaknesses and strengths in collaboration with robot," *Period. Polytech. Mech. Eng.,* vol. 52, no. 1, 25-33, 2008.

[32]   C. Lenz, S. Nair, M. Rickert, A. Knoll, W. Rösel, J. Gast, A. Bannat, and F. Wallhoff,"Joint-action for humans and industrial robots for assembly tasks," In: *Proc. 17th IEEE International Symposium on Robot and Human Interactive Communication,* Munich, 2008.

[33]   T. Wojtara, M. Uchihara, H. Murayama, S. Shimoda, S. Sakai, H. Fujimoto, and H. Kimura, "Human-robot collaboration in precise positioning of a three-dimensional object," *Automatica*, vol. 45, no. 2, 333-342, Feb. 2009.

[34]   M.P. Mayer, B. Odenthal, M. Faber, J. Neuhöfer, W. Kabuß, B. Kausch, and C.M. Schlick, "Cognitive engineering for direct human-robot cooperation in self-optimizing assembly cells," In: *1st International Conference on Human Centered Design*, (HCD 2009), San Diego, 2009, pp. 1003-1012.

[35]   N. Kim, D. Shin, R.A. Wysk, and L. Rothrock, "Using finite state automata (FSA) for formal modeling of affordances in human-machine cooperative manufacturing systems," *Int. J. Prod. Res.*, vol. 48, no. 5, 1303-1320, 2010.

[36]   J. Krüger, T.K. Lien, and A. Verl, "Cooperation of human and machines in assembly lines," *CIRP Ann. - Manuf. Technol.,* vol. 58, no. 2, 628-646, 2009.

[37]   D. Henrich, and S. Kuhn, "Modeling Intuitive behavior for safe human/robot coexistence cooperation," In: *Robotics and Automation, (ICRA 2006),* 2006, pp. 3929-3934.

[38]   M. Zaeh, and W. Roesel, "Safety aspects in a human-robot interaction scenario: A human worker is co-operating with an industrial robot," In: *12th FIRA RoboWorld Congress on Progress in Robotics*, Incheon: South Korea, 2009, pp. 53-62.

CHAPTER 7

Kinematics Analysis and Teleoperation of a 4-DOF Reconfigurable Modular Robot

Dan Zhang[*], Zhen Gao, Jianhe Lei and Zhanglei Song

Faculty of Engineering and Applied Science, University of Ontario Institute of Technology, 2000 Simcoe Street North, Oshawa, Ontario, L1H 7K4, Canada

Abstract: This research presents the kinematics analysis and the teleoperation system of a 4-DOF modular reconfigurable robot. The forward kinematic analysis of the designed robot is conducted firstly. A numerical method with the backpropagation neural network is investigated to solve inverse kinematics problem and the training samples for the neural network are obtained with FARO laser tracker, which guarantees modeling accuracy of the non-linear mapping from the task space to the joint space. The remote control system has a client-server structure. The local robot control computer plays a role as a server and remote terminal as a client. Simulation, monitoring and control of the robot can be conducted in a remote manner. When client works in the simulation mode, users can simulate robot motion with the virtual one. While the system works in the monitoring/control regime, the robot can be remotely controlled and the status of robot can be monitored dynamically at client site. By using Java3D technique, virtual model of the robot is implemented and only small data parcels with the current robot coordinates need to be transmitted to the visualization module of client side. This accelerates the system response and provides the operator with a synthesized view of the real robot.

Keywords: Kinematics analysis, teleoperation, reconfigurable robot, modular robot, java3d, remote control, forward kinematic, inverse kinematics, neural network, visualization module.

1. INTRODUCTION

As one important component in a rapidly reconfigurable robotic workcell, modular robot can be realized different function to meet the versatile need of today's manufacturing. Meanwhile, the reconfigurable modular robot also has many potential applications in the areas such as space, underwater, or other

---

*Address correspondence to Dan Zhang: Faculty of Engineering and Applied Science, University of Ontario Institute of Technology, 2000 Simcoe Street North, Oshawa, Ontario, L1H 7K4, Canada; Tel: 905.721.8668 ext. 5721; Fax: 905.721.3370; E-mail: Dan.Zhang@uoit.ca

complex and difficult environments [1-4].

Apart from reconfigurability of a robot, remote control can also vastly enrich the range of applications. With the rapid development of the Internet, it can act as a medium connecting many locations using one single protocol. The subject of robotic remote control has been widely studied and a survey can be found in [5], also describing an available public implementation of an Internet robot controller [6]. Another Web-interfaced, force-reflecting teleoperation systems can be found in [7]. One of the first industrial arms made available on the World Wide Web is presented in [8]. In [9, 10], two different real-time control systems of an Internet robot are descried.

Although there are some remote-operated robots available now, the authors have not encountered any available system that consider the issues, such as reconfigurability, real-time remote control and Java3D implementation of robot arm at the same time. Therefore, in this paper, we introduce a remote control system of a 4-DOF serial reconfigurable robot. Meanwhile, the forward/inverse kinematics analysis and Java3D visualization of the robot arm are also presented.

2. KINEMATIC MODELING

A. Forward Kinematic Analysis

Fig. 1 shows the geometric structure of the robot arm, respectively. Robot kinematics, which is classified as forward kinematics and inverse kinematics, is one of the fundamental issues in robot motion analysis and control.

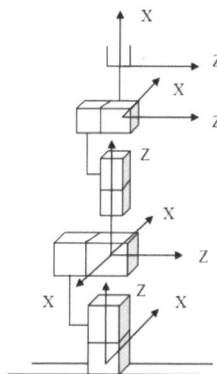

Figure 1: Analysis of the 4-DOF modular robot.

Table 1: The DH Table

i	a_i	b_i	α_i	θ_i
1	0	L_1	90°	θ_1
2	0	0	90°	θ_2
3	0	$L_3 + L_4$	90°	θ_3
4	L_5	0	0°	θ_4

Forward kinematics is the computation of the position and orientation of robot's end-effector as a function of its joint angles. Using Denavit-Hartenberg method, D-H table of 4-DOF modular robot can be shown in Table **1**, where L_i is the distance between each joint. The angle between X_i and X_{i+1} is defined as θ_i. The distance between Z_i and Z_{i+1} is defined as a_i. The absolute value of b_i is the distance between X_i and X_{i+1}, the angle between Z_i and $Z_{i+1}$is defined as α_i. With the D-H table and the rotation matrix, position vector p of the end-effector can be calculated as follow:

.. (1)

$$a_i = \begin{bmatrix} a_i \cos\theta_i \\ a_i \sin\theta_i \\ b_i \end{bmatrix}$$

(2)

$$p = a_1 + Q_1 a_2 + Q_1 Q_2 a_3 + Q_1 Q_2 Q_3 a_4$$

(3)

$$p = \begin{bmatrix} (c\theta_1 s\theta_2 + (c\theta_1 c\theta_2 - s\theta_1)s\theta_3)L_5 c\theta_4 \\ (s\theta_1 s\theta_2 + (s\theta_1 c\theta_2 - c\theta_1)s\theta_3)L_5 c\theta_4 \\ L_1 - c\theta_2(L_3 + L_4) + s\theta_2 c\theta_3 c\theta_4 L_5 - c\theta_2 s\theta_4 L_5 \end{bmatrix}$$

(4)

Where $s\theta_i = \sin(\theta_i)$ and $c\theta_i = \cos(\theta_i)$, respectively.

Thus, the reachable workspace of the 4-DOF modular robot can be derived based on the forward kinematic solution. As shown in Fig. **2**, its workspace is a hollow sphere.

Figure 2: The reachable workspace of the 4-DOF modular robot.

B. Inverse Kinematic Analysis

Artificial neural networks (ANN) are massively parallel adaptive networks of simple nonlinear computing elements called neurons. The main function of neural networks is establishing the complex nonlinear relationship between inputs and outputs without deducing the mathematics expression. ANN is used for solving Inverse kinematic problem (IKP) in the sense that they are able to create internal representations through training example sets. And also mathematical solutions for inverse kinematics problems may not always correspond to physical solutions and method of its solution depends on the robot configuration. For the 4-DOF robot arm, it even can be set up in nine different configurations. Therefore, we investigated to solve IKP using ANN method.

Multilayer perceptron (MLP) is the most widely used neural network. A graphical representation of an MLP for the solution of IKP is shown in Fig. **3**. The optimal number of hidden neurons is set to 30 by trial and error. The neurons at hidden layers have sigmoid activation functions. Training procedure adjusted the weighting coefficients using back-propagation algorithm [11].

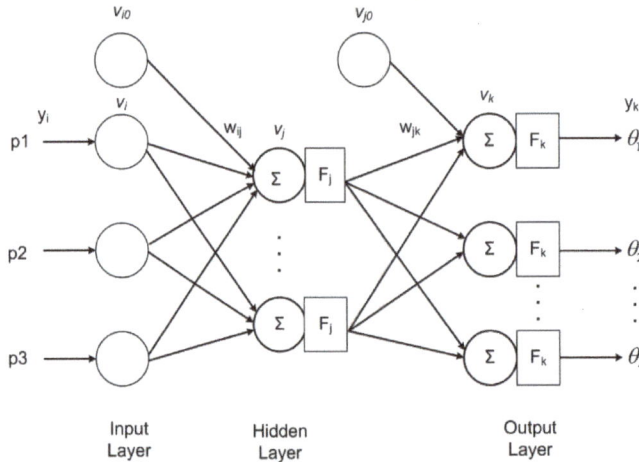

Figure 3: Structure of an artificial neural network.

Training data set was obtained in this way: with the software we programmed, joint angles θ1, θ2, θ3 and θ4 are set into different values, and then we used FARO laser tracker to measure a series of coordinates x, y and z of the end-effector in Cartesian space (Figs. **4** and **5**). FARO Laser Tracker Model ION can measure 3D coordinates with laser beam by following a mirrored spherical probe (MSP). The tracker can follow MSP in range of 25m, tracking features at 10000 points per second with accuracy of $8\,\mu$ m + 0.4 μ m/m. 500 data pairs were obtained finally. 20% of the vectors are used to validate how well the network, the last 20% of the vectors provide an independent test of network generalization to data that the network has never seen [12].

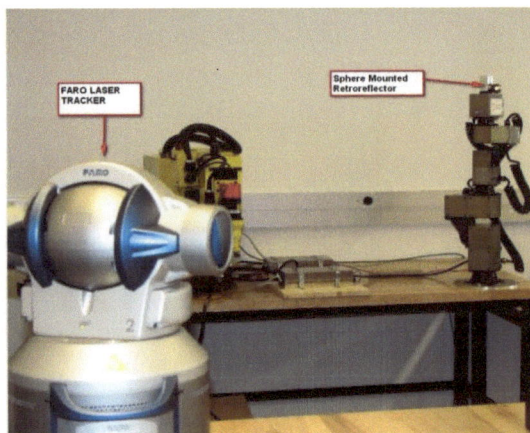

Figure 4: Laser tracker and robot arm set up.

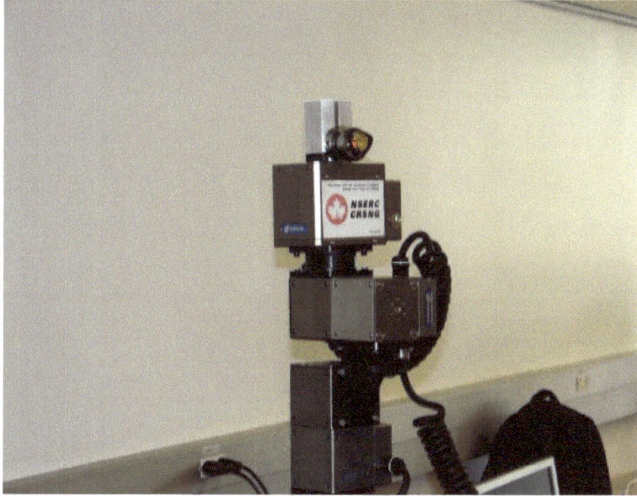

Figure 5: Sphere mounted retroreflector.

A possible use of the modular reconfigurable robot is surface polishing or sanding. To show this a model of a belt sander has been created and attached to the end-effector. Two sides of the belt sander have a slack area for sanding and the other has a solid plate for sanding. The model of this concept can be seen in Fig. **6**.

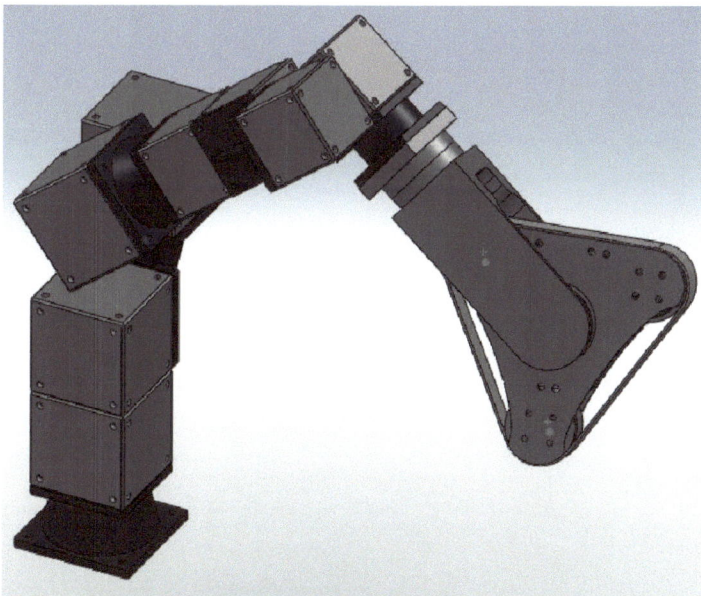

Figure 6: The representative application of the reconfigurable modular robot.

3. FINITE ELEMENT ANALYSIS

An FEA was done on the working configuration used under the conditions of the proposed deburring operation. The bottom PAM connector has a fixed constraint applied to it and the end effector has static loads applied to result in solutions (see Fig. **7**). In this case a load of 20 lbf was applied to the end effector. Displacement, elemental stress and reaction force will be investigated.

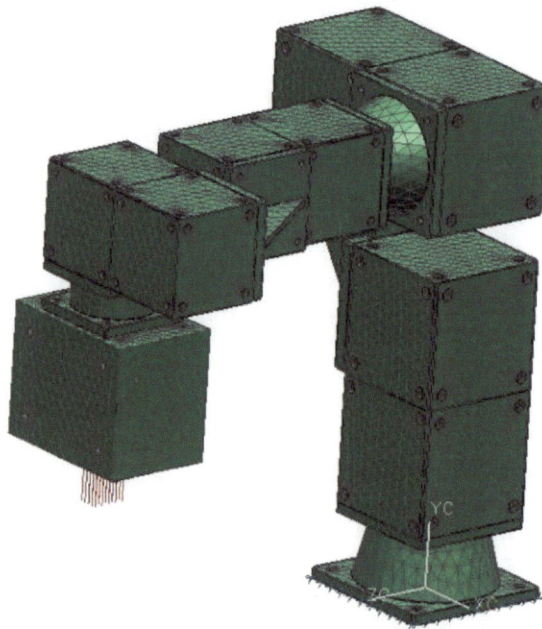

Figure 7: 3D Tetrahedral Mesh for Work Pose.

Table 2: Work Pose Results

Load Case 1: 20 lbf								
	Displacement (mm)				Stress (mN/mm^2(kPa))			
	X	Y	Z	Magnitude	Von-Mises	Min Principal	Max Principal	Max Shear
Static Step 1								
Max	9.38E-03	5.97E-02	2.21E-02	6.35E-02	3.27E+03	5.89E+02	3.47E+03	1.65E+03
Min	-6.11E-03	-5.59E-03	-2.24E-02	0.00E+00	1.83E-01	-3.04E+03	-3.14E+02	1.05E-01

As shown in Table **2**, the solution to the finite element analysis with 20 lbf provides us with the maximum displacement being 6.35 E-2 mm.

FEA was conducted to analyze break points, base constraint needs and largest unwanted movement when a load is applied to the end effector. As shown in Figs. **9** and **10**, in the work pose the break point is in the cylindrical connector and the largest displacement is seen in the end effector.

4. REMOTE CONTROL SYSTEM DESIGN

A. General Description of the Control System

There are three main parts in the system - client, server and communication network (Fig. **8**). Clients can run at the remote place while the server runs at robot location, data communication is realized by the CAN Bus (local control) and internet (remote control).

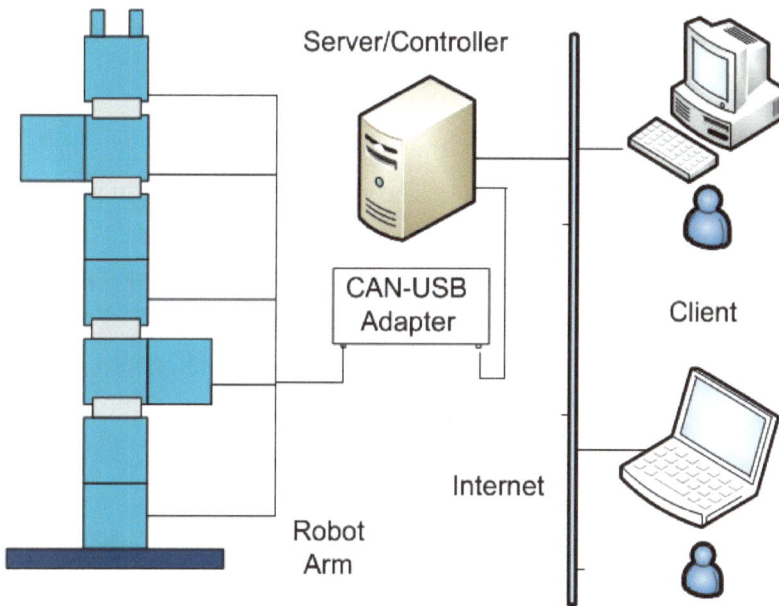

Figure 8: System hardware architecture.

The system is capable of operating in two control modes: simulation and monitoring/control. In the monitoring/control regime, the operator remotely controls the real robot, while in the simulation mode, he or she controls the virtual robot, *i.e.* a 3D graphical model of the modular robot. In the latter regime, work is performed fully on the client part of the system, which is useful for preliminary testing of the operation prior to operating with the real equipment.

B. Development of Server Control Platform

1) Hardwares and Local Control Network: The hardware of the local control system consists of a host computer (server), CAN Bus-USB adapter, the CAN Bus distributed subsystem and the reconfigurable modular robot. There are a number of configurations to build the robotic arm in order to reach different workspaces. The 4-DOF robotic arm is assembled as a part of remote control of reconfigurable robotic system. It can be reconfigured into nine different forms. The robotic arm is built by PowerCube modules. These modules can be connected using the communication protocols, such as RS-232, RPROFIBUS-DP or CAN Bus.

In this system, CAN Bus is chosen to connect internal controllers because CAN protocol is a fast serial bus that is designed to provide an efficient, reliable and very economical link between sensors and actuators.

The master control system generates the sequential program and stores the current sequential commands in the module. The subsequent command is stored in the buffer, and then the command is sent to the connected modules step by step. The power system is set up according to total modules voltage and current. It supplies both power voltage and logic voltage in order to make all the modules work properly. The communication between robotic arms and PC controller is connected by the CAN bus. One terminal is connected *via* 9-pin SUB-D port on the terminal block and the other terminal is connected *via* USB bus on PC. The maximum data transfer rate is 1 Mbit/s, which is constructed and transmitted by the CAN chip. The CAN protocol corresponds to the data link layer in the ISO (International Organization for Standardization) and OSI (Open System Interconnection Reference Model) reference model, which meets the real-time requirements [13].

2) Software Design at Server Side: The server control panel is designed to be programmed with Visual Basic.net. VB.net is not only a language, but also primarily an integrated, interactive development environment (IDE) [14]. The VB.net-IDE has been highly optimized to support rapid application development. There are three modules on the server side: a connection control module, a local controller module and a data transaction module.

The connection control module is mainly used for controlling the connection from client side. It offers the following functions:

i) Listening to a connection and disconnection request from client.

ii) Queuing the connection request in a waiting list when server is busy.

iii) Removing the cancelled connection request or releasing the invalid connection.

iv) Receiving command from client and transmitting to local controller module.

The local controller module is mainly used for operating the real robotic arm. It provides the following functions:

i) Connecting to the real modular robotic arm *via* CAN bus connection.

ii) Receiving commands from the local server controller or remote controller

iii) Executing it on the modular robotic arm *via* CAN bus connection.

The data transaction module takes the responsibility for validating and executing the robot controlling language program. These applications programming Interface (API) can be classified into four groups:

i) CAN communication setup. This function group is used to open the CAN communication ports by using the initialization string (initstring) and to establish the communication among CAN nodes of the robot inside nodes.

ii) Reading and writing. This function group is used to transfer important information of the robot arm such as position, speed, acceleration, consumption power, response coefficients to control the motor, *etc.*

iii) Movement control. This function group gives the user the capacity to indicate the final position, speed and acceleration of the arm elements.

iv) Input/output control. This function group is used to control the I/O of the modules. The structure and classification of the functions are based on the library m5apw32.dll which has the prototype. The return value of the function indicates if the function has been or has not been executed.

C. Development of Client Control Platform

1) Software Design at Client Side: The software at client side contains three modules: a connection control module, a 3D robot model module, and a data transaction module. It offers the following functions:

(a) Sending connection request to server.

(b) Sending disconnection request to server.

(c) Keeping the connection.

(d) Sending users inputted program to server.

The 3D model module provides a view of 3D virtual robot model and offers the following functions:

(a) Displaying the 3D model of robotic arm in the virtual environment.

(b) Receiving the operation message from the program editor.

(c) Operating the motion according to the received control message.

The data transaction module takes the responsibility for validating and executing the robot-controlling language program. It has the following functions:

(a) Allowing users to input data and calculate the inverse kinematics.

(b) Checking error of the input.

(c) Applying the valid operation code to the 3D robotic arm model and simulating at the client side.

(d) Transferring the operation message to server side.

2) Client Control GUI: The graphic user interface of client side should include an operation window and a 3D robot model window. In the corresponding textbox, users can input the rotation angle directly or input the position coordinate and solved by inverse kinematic by clicking the calculation button. Once users retrieve the rotational angle for each module, they can simulate on the Java3D model by clicking the execute button and/or send the data to server side to control the robotic arm by clicking the send button. Reset button can be used to clear all the current status. The action button controls the 2 finger gripper directly and instructs it to perform picking, holding and releasing actions. The Exit button allows users to close the connection and exit the program safely.

D. Remote Monitoring and Control

The software for communicating over the Internet is implemented with Socket mechanism. There are several communication protocols in the world such as User Datagram Protocol (UDP), Transmission Control Protocol (TCP) *etc.* Besides, for every client registered with the Registrar, there exists one assigned Subscriber who is knowledgeable of the user's request and responsible of maintaining an active channel between the client and the server. Multi-threading techniques are adopted for implementing the sensor data processing modules. An authorized user can control a remote device, indirectly, through the Control Commander. Fig. **9** illustrates the data processing mechanism for real-time monitoring and control.

5.   VISUALIZATION   AND   IMPLEMENTATION   OF   THE RECONFIGURABLE MODULAR ROBOT

Pick and place operation is one of the typical tasks to take a workpiece from a given initial pose to final pose specified by the position and orientation in certain coordinate frame. A picking and placing experiment, as shown in Fig. **10** is implemented based the integration of the aforementioned remote control approach and visualization method. Table **3** gives the obtained experimental results.

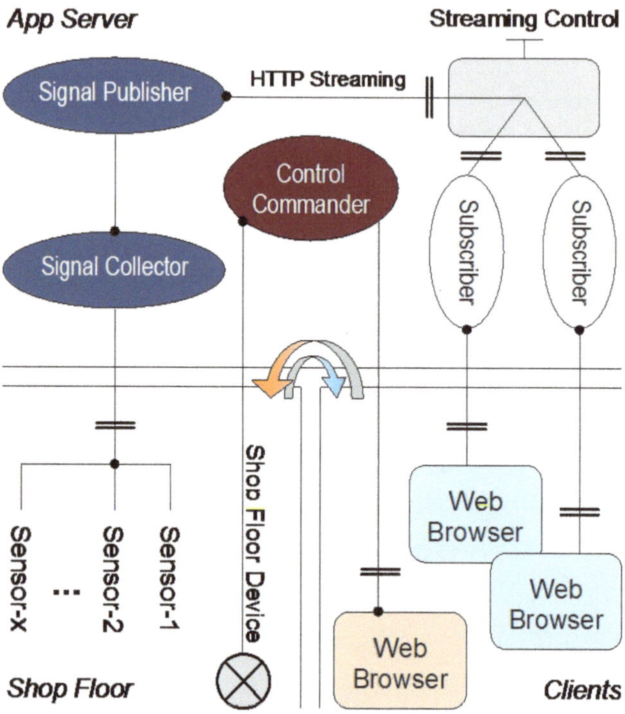

Figure 9: Monitoring and control.

Figure 10: Test on picking and placing motion.

Table 3: Experimental Data

Pose	θ_1	θ_2	θ_3	θ_4	*Vel*	*Acc*
Initial	-45°	95°	0	-85°	0	0
Rising	-50°	90°	0	-80°	0.15*m/s*	0.6*m/s²*
Falling	-135°	75°	0	-65°	0.15*m/s*	0.6*m/s²*
Final	-150°	80°	0	-70°	0	0

6. CONCLUSIONS

This research focuses on kinematic analysis and remote control of the 4-DOF reconfigurable modular robot. A new method based on the artificial neural network using backpropagation algorithm is investigated to solve the inverse kinematics problem. The proposed remote control system has the simulation, monitoring and control functions. Virtual model of the robot is implemented with Java3D and only small data parcels with the current robot coordinates need to be transmitted to the visualization module of client side. This accelerates the system response and provides the operator with a synthesized view of the real robot.

ACKNOWLEDGEMENT

Declared none.

CONFLICT OF INTEREST

The author(s) confirm that this chapter content has no conflict of interest.

REFERENCES

[1]   I. Chen, "Rapid response manufacturing through a rapidly reconfigurable robotic workcell," *Robot. Comput.-Integr. Manuf.*, vol. 17, no. 3, pp. 199-213, 2001.
[2]   F. Aghili and K. Parsa, "A reconfigurable robot with lockable cylindrical joints," *IEEE Trans. Robot.*, vol. 25, no. 4, pp. 785-797, 2009.
[3]   S. Murata, E. Yoshida, A. Kamimura, H. Kurokawa, K. Tomita, and S. Kokaji, "M-TRAN: self-reconfigurable modular robotic system," *IEEE/ASME Trans. Mechatron.*, vol. 7, no. 4, pp. 431-441, 2002.
[4]   Y. Liu and G. Liu, "Track--stair interaction analysis and online tipover prediction for a self-reconfigurable tracked mobile robot climbing stairs," *IEEE/ASME Trans. Mechatron.*, vol. 14, no. 5, pp. 528-538, 2009.

[5]   K. Taylor, B. Dalton, and J. Trevelyan, "Web-based telerobotics," *Robotica*, vol. 17, no. 1, pp. 49-57, 1999.

[6]   C. Wronka and M. Dunnigan, "Internet remote control interface for a multipurpose robotic arm," *Int. J. Adv. Robot. Syst.,* vol. 3, pp. 179-182, 2006.

[7]   R. Oboe, "Web-interfaced, force-reflecting teleoperation systems," *IEEE Trans. Ind. Electron.,* vol. 48, no. 6, pp. 1257-1265, 2001.

[8]   K. Goldberg, M. Mascha, S. Gentner, N. Rothenberg, C. Sutter, and J. Wiegley, "Desktop teleoperation *via* the world wide web," *in IEEE International Conference on Robotics and Automation*, Citeseer, 1995, pp. 654-654.

[9]   I. Elhaij, H. Hummert, N. Xi, W. Fung, and Y. Liu, "Real time bilateral control of internet based teleoperation". *in Proceedings of the 3rd World Congress on Intelligent Control and Automation*, Hefei, China, 2000, pp. 3761-3766.

[10]   J-H. Park, J. Park, and S. Moon, "Real-time bilateral control for an internet-based telerobotic system," *JSME Int. J. Ser. C*, vol. 47, no. 2, pp. 708-714, 2004.

[11]   J. Lei and Y. Ge, "Application of neural network to nonlinear static decoupling of robot wrist force sensor," *in Intelligent Control and Automation, the Sixth World Congress on*, 2006, pp. 5282-5285.

[12]   H. Demuth, M. Beale, and M. Hagan, *Neural Network Toolbox 6*, The MathWorks, 2008

[13]   K. Tindell, H. Hansson, and A. Wellings, "Analysing real-time communications: Controller area network (CAN)," *in Proceedings 15th IEEE Real-Time Systems Symposium*, Citeseer, 1994. pp. 259-265.

[14]   F. Balena, *Programming Microsoft Visual Basic .NET (Core Reference)*. Microsoft Press Redmond, WA, USA. 2002.

INDEX

Human finger motion 37, 39
Human operators 141, 144, 146-7, 153-8
Human-robot collaboration 151-2, 156
  direct 156-7
Human-robot interaction 140
Humanoid robot 17, 21, 29
  low-cost 10-11, 16, 40
Humanoid robot iCub 20-2, 30, 33

I

i-th phalanx 13
iCub 29
iCub robot 28, 40
Industrial robots 88, 103, 105, 152
Influence of Joint Vector 94, 98
Information models 67
  large 67
Intelligent assembly 140
Interactive development environment (IDE) 170
Internet robot controllers 163
Interoperability & traceability 63
Interpolating cubic lines 39
Inverse kinematic problem (IKP) 162, 165
Inverse kinematics 87, 92, 111, 162-3, 172-3
Inverse kinematics solution 89, 114

J

Joint structures 21-2
Joint vector 86, 90, 94, 98, 100

K

Kinematic characteristics 8, 10
Kinematic configuration 105-6
Kinematic parameters 8
Kinematic scheme 9-10
Kinematic simulation 10-11
Kinematic Structure 102, 108-9, 126, 133
Kinematics 4-5, 15, 23-4, 37, 102, 104-5, 107-8, 133
Kinematics analysis 162

L

M